# NAVAL FORCES' CAPABILITY FOR THEATER MISSILE DEFENSE

Committee for Naval Forces' Capability for Theater Missile Defense
Naval Studies Board
Division on Engineering and Physical Sciences
National Research Council

NATIONAL ACADEMY PRESS
Washington, D.C.

**NATIONAL ACADEMY PRESS • 2101 Constitution Ave., N.W. • Washington, DC 20418**

NOTICE: The project that is the subject of this report was approved by the Governing Board of the National Research Council, whose members are drawn from the councils of the National Academy of Sciences, the National Academy of Engineering, and the Institute of Medicine. The members of the committee responsible for the report were chosen for their special competences and with regard for appropriate balance.

This work was performed under Department of the Navy Contract N00014-00-G-0230/0003 issued by the Office of Naval Research under contract authority NR 201-124. However, the content does not necessarily reflect the position or the policy of the Department of the Navy or the government, and no official endorsement should be inferred.



Cover: Photographs courtesy of the U.S. Navy.

International Standard Book Number 0-309-07426-6



Copies available from:

Naval Studies Board
National Research Council
2101 Constitution Avenue, N.W.
Washington, DC 20418

Printed in the United States of America

# THE NATIONAL ACADEMIES

National Academy of Sciences
National Academy of Engineering
Institute of Medicine
National Research Council

The **National Academy of Sciences** is a private, nonprofit, self-perpetuating society of distinguished scholars engaged in scientific and engineering research, dedicated to the furtherance of science and technology and to their use for the general welfare. Upon the authority of the charter granted to it by the Congress in 1863, the Academy has a mandate that requires it to advise the federal government on scientific and technical matters. Dr. Bruce M. Alberts is president of the National Academy of Sciences.

The **National Academy of Engineering** was established in 1964, under the charter of the National Academy of Sciences, as a parallel organization of outstanding engineers. It is autonomous in its administration and in the selection of its members, sharing with the National Academy of Sciences the responsibility for advising the federal government. The National Academy of Engineering also sponsors engineering programs aimed at meeting national needs, encourages education and research, and recognizes the superior achievements of engineers. Dr. William A. Wulf is president of the National Academy of Engineering.

The **Institute of Medicine** was established in 1970 by the National Academy of Sciences to secure the services of eminent members of appropriate professions in the examination of policy matters pertaining to the health of the public. The Institute acts under the responsibility given to the National Academy of Sciences by its congressional charter to be an adviser to the federal government and, upon its own initiative, to identify issues of medical care, research, and education. Dr. Kenneth I. Shine is president of the Institute of Medicine.

The **National Research Council** was organized by the National Academy of Sciences in 1916 to associate the broad community of science and technology with the Academy's purposes of furthering knowledge and advising the federal government. Functioning in accordance with general policies determined by the Academy, the Council has become the principal operating agency of both the National Academy of Sciences and the National Academy of Engineering in providing services to the government, the public, and the scientific and engineering communities. The Council is administered jointly by both Academies and the Institute of Medicine. Dr. Bruce M. Alberts and Dr. William A. Wulf are chairman and vice chairman, respectively, of the National Research Council.

**COMMITTEE FOR NAVAL FORCES' CAPABILITY FOR THEATER MISSILE DEFENSE**

ALAN BERMAN, Applied Research Laboratory, Pennsylvania State University, *Chair*
ROY R. BUEHLER, Mableton, Georgia
WILLIAM A. DAVIS, JR., Guntersville, Alabama
LARRY E. DRUFFEL, South Carolina Research Authority
BRIG "CHIP" ELLIOTT, BBN Technologies
FRANK A. HORRIGAN, Raytheon Company (retired)
RICHARD J. IVANETICH, Institute for Defense Analyses
HARRY W. JENKINS, ITT Industries
DAVID V. KALBAUGH, Applied Physics Laboratory, Johns Hopkins University
FRANK KENDALL, Falls Church, Virginia
L. DAVID MONTAGUE, Menlo Park, California
F. ROBERT NAKA, CERA, Inc.
J. THEODORE PARKER, Annapolis, Maryland
C. KUMAR N. PATEL, Pranalytica, Inc.
NILS R. SANDELL, JR., ALPHATECH, Inc.
HOWARD E. SHROBE, Massachusetts Institute of Technology
JOHN P. STENBIT, TRW, Inc.
ROBERT F. STENGEL, Princeton University
EDWARD J. WEGMAN, George Mason University
STEPHEN D. WEINER, Lincoln Laboratory, Massachusetts Institute of Technology

*Staff*

CHARLES F. DRAPER, Study Director
SIDNEY G. REED, JR., Consultant
JAMES G. WILSON, Consultant

## NAVAL STUDIES BOARD

*Marine Corps Liaison Representative*

LTGEN JOHN E. RHODES, USMC, Commanding General, Marine Corps Combat Development Command (through August 17, 2000)
LTGEN BRUCE B. KNUTSON, JR., USMC, Commanding General, Marine Corps Combat Development Command (as of August 18, 2000)

RONALD D. TAYLOR, Director
CHARLES F. DRAPER, Senior Program Officer
MARY G. GORDON, Information Officer
SUSAN G. CAMPBELL, Administrative Assistant
KERRY A.M. WILLIAMS, Research Assistant

# Preface

Under the precepts of *Joint Vision 2020,*[1] full-dimensional protection is one of four principal operational concepts that describe future war-fighting objectives. Defense against cruise and ballistic missiles remains a key challenge for the military forces. The missile threat to these forces continues to grow with the continued global diffusion of missile technologies and the expansion of access to space-based reconnaissance and imagery. The threat is made even more difficult to defeat if the missile warheads are armed with chemical or biological warfare payloads. The goal of the Department of Defense is to develop an integrated, in-depth theater air and missile defense system that exploits capabilities to detect, identify, locate, track, and deny enemy missile attacks on joint forces and friendly nations. For naval forces operating in a littoral environment the cruise missile defense problem is further complicated by the need to detect, identify, and track cruise missiles against the land clutter background.

The Navy's Aegis system was developed as part of a layered integrated system of ship-based sensors and weapons to protect ships against air- and ground-launched cruise missile attack. The effectiveness of current Navy and Marine Corps cruise missile defenses should be enhanced by the planned introduction of the cooperative engagement capability (CEC), which will allow participants in a CEC network to contribute to the development of target tracks based on detections by geographically dispersed sensors.

---

[1]Shelton, GEN Henry H., USA. 2000. *Joint Vision 2020.* Joint Chiefs of Staff, The Pentagon, Washington, D.C. Available online at <http://www.dtic.mil/jv2020/jvpub2.htm>.

The Navy and the other Services are developing systems that can contribute to theater ballistic missile defense. Ultimately, theater ballistic missile defense systems will evolve from the layered capabilities of several independent systems. These systems might include the Patriot advanced capability-3 (PAC-3), the medium extended air defense system (MEADS), the theater high altitude area defense (THAAD) system, the airborne laser (ABL), the Navy area defense (NAD) system, and the Navy theater wide (NTW) system. Navy programs must be evaluated in the context of the capabilities and likely availability of systems fielded by other Services and in the context of the mutual integration of Navy/ Marine Corps and the other Services' systems.

As currently envisioned, naval forces will rely on evolved variants of the Aegis system. These will employ an improved standard missile (SM) to provide an in-theater (at sea) capability to engage ballistic missiles within the atmosphere (the NAD system) and an advanced SM to engage ballistic missiles at longer ranges outside the atmosphere (the NTW system). Forward-deployed naval forces would make this contribution in developing phases of conflict, protecting threatened nations and arriving joint forces against attacks by ballistic missiles that can deliver weapons of mass destruction.

The development of a robust theater missile defense capability will demand technological advances in a number of areas. Based on current concepts, the successful engagement of attacking ballistic missiles will depend on the availability of effective hit-to-kill interceptors, multispectral seekers, and improvements to radar performance that will provide an ability to detect, evaluate, and overcome penetration aids and other countermeasures to theater missile defense. The successful engagement of hostile cruise missiles in a littoral environment will depend on advances in radar performance in a high-clutter background; networked, distributed, surveillance capabilities; low-observable detection technology; and data processing and fusion. The network that integrates the layered capabilities will require very low latencies and high bandwidth so that data and information are delivered when and where needed.

The committee was asked to evaluate all of these factors in the context of the Department of the Navy's current and planned acquisitions and in the context of its current investment in research and development (R&D).

## TERMS OF REFERENCE

At the request of Admiral Jay L. Johnson, USN, former Chief of Naval Operations, the National Research Council, under the auspices of the Naval Studies Board, conducted a study of current and future naval theater missile defense capabilities. The terms of reference for the study are as follows:

- Evaluate present and projected future ballistic and cruise missile threats to naval forces operating in littoral areas and to joint force operations in these areas.

- Evaluate the current state of technologies involved in theater missile defense, accounting for the efforts of the other Services and defense agencies. Project (out to 2015) the future state of the technologies involved.
- Evaluate current and projected Department of the Navy programs designed to meet the threats. Evaluate current and projected R&D programs aimed at providing naval forces with new and improved capabilities, including the Navy's own efforts and those of other Services.
- Recommend R&D priorities, accounting for the potential technical and operational interactions among Navy, other Services and defense agencies, and allied nations' programs.
- The assessment should include consideration of existing and planned platform, missile, and command, control, communications, computing, intelligence, surveillance, and reconnaissance (C4ISR) systems and the capability of these systems to support the ability of the naval forces to contribute to the development of a robust theater missile defense for naval forces in expeditionary operations and to cooperatively protect joint forces in joint operations.[2]

In a letter dated December 11, 2000, to the president of the National Academy of Sciences, General James L. Jones, USMC, Commandant of the Marine Corps, indicated that he also endorsed the study's terms of reference.

## COMMITTEE MEETINGS

The Committee for Naval Forces' Capability for Theater Missile Defense first convened in April 2000 and held further meetings and site visits over a period of 8 months:

- April 25-26, 2000, in Washington, D.C. Organizational meeting: Navy and Marine Corps briefings on operational requirements and naval missile defense programs and Office of Naval Research and Naval Research Laboratory briefings on naval theater missile defense science and technology efforts.
- May 23-24, 2000, in Washington, D.C. Central Intelligence Agency, Defense Intelligence Agency, and Office of Naval Intelligence briefings on missile threats; Ballistic Missile Defense Organization (BMDO), Joint Theater Air Missile Defense Organization, Deputy Assistant Secretary of the Navy for Theater Combat Systems, Naval Sea Systems Command (PMS 456), Program Executive Office for Theater Surface Combatants (PEO TSC), Army Program Executive Office for Air and Missile Defense, and Air Force Global Power Program Office

[2]Adopting a usage common in the briefings to the committee, the report uses the term "battle management command, control, and communications" (BMC3) to encompass things that could also be described as C4ISR.

briefings on theater missile defense programs, operations, and technologies; and Center for Naval Analyses briefing on the limitations of the Anti-Ballistic Missile Treaty.

- June 27-28, 2000, in Washington, D.C. PEO TSC and Program Executive Office for Expeditionary Warfare briefings on NAD and NTW systems; Naval Sea Systems Command (PMS 500) and Raytheon Electronic Systems briefing on MFR/DD21; and BMDO system engineering briefing.
- July 25-26, 2000, in Washington, D.C. PEO TSC, Joint Technology Program, Naval Sea Systems Command (PMS 467), Space and Naval Warfare Systems Command (PMW 159), and Joint Land Attack Cruise Missile Defense Elevated Netted Sensors Program Office briefings on missile defense programs, operations, and technologies.
- August 29-30, 2000, in Washington, D.C. Deputy Commandant for Aviation and Office of the Chief of Naval Operations (N865) briefings on naval theater missile defense operational requirements and programs; U.S. Joint Forces Command briefing on theater missile defense documentation; Applied Research Laboratory (Pennsylvania State University), Army Space and Missile Defense Command, and Joint Non-Lethal Weapons Directorate briefings on advanced electro-optics and laser systems; and Institute for Defense Analyses and Naval Sea Systems Command briefing on single integrated air picture efforts.
- August 31, 2000, in Laurel, Maryland. Small group site visit to the Applied Physics Laboratory (Johns Hopkins University) to follow up on information presented on the NAD and NTW systems presented at the June 27-28 meeting.
- September 8, 2000, in Lexington, Massachusetts. Small group site visit to Lincoln Laboratory (Massachusetts Institute of Technology) for briefings on radar/infrared discrimination, SPY radar, open architecture, and combat identification techniques.
- September 11-15, 2000, in Woods Hole, Massachusetts. Committee deliberations and report drafting.
- November 28-29, 2000, in Washington, D.C. Committee deliberations and report drafting.

The months between the last meeting and publication of the report were spent preparing the draft manuscript, reviewing and responding to the external review comments, and editing the report.

# Acknowledgment of Reviewers

This report has been reviewed in draft form by individuals chosen for their diverse perspectives and technical expertise, in accordance with procedures approved by the National Research Council's (NRC's) Report Review Committee. The purpose of this independent review is to provide candid and critical comments that will assist the institution in making the published report as sound as possible and to ensure that the report meets institutional standards for objectivity, evidence, and responsiveness to the study charge. The review comments and draft manuscript remain confidential to protect the integrity of the deliberative process. We wish to thank the following individuals for their review of this report:

Gregory H. Canavan, Los Alamos National Laboratory,
Anthony J. DeMaria, DeMaria ElectroOptics,
James H. Doyle, Bethesda, Maryland,
Eugene Fox, McLean, Virginia,
Marvin J. Langston, SALUS Company, Inc.,
George Lewis, Massachusetts Institute of Technology,
Donald L. Pilling, Battelle Memorial Institute,
Merrill I. Skolnik, Baltimore, Maryland, and
Keith A. Smith, Vienna, Virginia.

Although the reviewers listed above provided many constructive comments and suggestions, they were not asked to endorse the conclusions and recommendations, nor did they see the final draft of the report before its release. The review of this report was overseen by Robert A. Frosch, appointed by the NRC's Report Review Committee, who was responsible for making certain that an independent examination of this report was carried out in accordance with institutional procedures and that all review comments were carefully considered. Responsibility for the final content of this report rests solely with the authoring committee and the institution.

# Contents

# Executive Summary

At the request of the Chief of Naval Operations, the National Research Council, under the auspices of the Naval Studies Board, established a committee to assess the Department of the Navy's current and future naval theater missile defense (TMD) capabilities. The Committee for Naval Forces' Capability for Theater Missile Defense first convened in April 2000 and met approximately 2 days a month for 8 months. This report is based on the information presented to the committee during that period and on the committee members' accumulated experience and expertise in military operations, systems, and technologies.

## ES.1 TODAY'S FRAMEWORK IN PERSPECTIVE

Through their evolving strategies Forward...From the Sea[1] and Operational Maneuver From the Sea (OMFTS),[2] the Navy and Marine Corps have acknowledged a shift in warfare from operations on the open seas to operations in and adjacent to littoral areas. This shift in warfare location presents many technical and operational challenges to naval forces in power projection, the most notable

[1]Department of the Navy. 1994. "Forward...From the Sea, Continuing the Preparation of the Naval Services for the 21st Century," U.S. Government Printing Office, Washington, D.C., September 19.

[2]Headquarters, U.S. Marine Corps. 1996. "Operational Maneuver From the Sea," U.S. Government Printing Office, Washington, D.C., January 4. Available online at <http://www.192.156.75.102/omfts.htm>.

TABLE ES.1 Future Naval Force Capabilities for Handling Cruise and Ballistic Missile Threats

| Threat | Capability |
|---|---|
| Antiship cruise missile (ASCM) | Multifunction radar (MFR)<br>Ship self-defense system (SSDS)<br>Evolved sea sparrow missile (ESSM)<br>SPY-1D(V) radar |
| Overland cruise missile (OCM) | E-2C Radar Modernization Program/AMTI radar with ADS-18 antenna (funding uncertain)<br>Complementary low altitude weapon system (CLAWS) |
| Ballistic missile | Navy area defense (NAD) system<br>Navy theater wide (NTW) system |

of which may be an increase in the land-based threat to the forces engaged in such operations.

Both theater ballistic missile defense (TBMD) and cruise missile defense (CMD; including antiship cruise missile defense (ASCMD) and overland cruise missile defense (OCMD)) are important emerging military capabilities that are inherently necessary if naval forces are to execute missions in littoral areas. Today, there are large numbers and varieties of cruise and ballistic missiles in the operational inventories of many potential future adversaries of the United States.

Although high-performance ballistic missiles exist and could become available to potential future adversaries, most of the ballistic missiles that are currently available to such adversaries are of rather unsophisticated design. Many have limited accuracy of delivery and are ineffective for hitting tactical targets. As currently configured, many are nonseparating, single-stage rockets that are less stressing to defense systems than are multistage missiles. Many others are not able to deploy penetration aids. In a military sense, these threats will have limited tactical value unless they carry nuclear, chemical, and/or biological warheads. However, even as currently configured, they pose a serious threat to deployed forces and assets, as well as to the political stability of neighboring or allied countries.

Future naval force capabilities for handling cruise and ballistic missile threats are shown in Table ES.1. Based on its assessment of these future capabilities and the evolving threat, the committee's conclusions can be summarized as follows:

- ASCMD, OCMD, and TBMD are essential for littoral operations. The threats to naval (and joint) forces operating in littoral areas stress the capabilities of current ASCMD, OCMD, and projected TBMD systems. All indications are that cruise and ballistic threats will become more stressing.

- Current ASCMD systems have marginal or poor performance in littoral areas against some existing advanced antiship cruise missile (ASCM) threats. The Navy has many significant improvements under development—e.g., multifunction radar (MFR), ship self-defense system (SSDS), evolved sea sparrow missile (ESSM) system, and SPY-1D(V) radar—which should be fielded as soon as possible. Some needed components are not under development (e.g., an ESSM launcher for non-Aegis combatants). Furthermore, naval combatants need an elevated detection platform and an over-the-horizon engagement system to restore an area defense capability providing the depth of fire needed for robust defense.
- The Navy area defense (NAD) and Navy theater wide (NTW) Block I systems will enable defeating some current unsophisticated ballistic missile threats; however, until upgraded systems are fielded, these systems will have limited capabilities against postulated advanced ballistic missile threats.
- The SM-2 Block IVA and SM-3 weapon programs associated with NAD and NTW Block I are well structured, but upgrades are required to the SPY-1 radar to make its capabilities compatible with the reach of the SM-3.
- Although both the NAD and NTW systems are based on the concept of spiral development (build-improve-build-improve . . .), the research and development (R&D) to support such a development concept is not in place.
- Negation of stressing overland cruise missile (OCM) and ASCM threats will require the Navy and Marine Corps to field new sensor and weapon capabilities and/or to become dependent on and integrated with nonorganic sensor systems of other Services and agencies.
- Naval forces lack a competent battle management command, control, and communications (BMC3) capability in terms of both concepts of operation and system effectiveness for missile defense in coordination with operations for offense in littoral areas. Inadequate procedures and technical capabilities exist for coordinating assets in the battle space, and current enhancement efforts are often based on legacy technology (e.g., Link 16) that does not support the necessary flexible modes of operation.

## ES.2 PRIORITIZATION OF CRUISE AND BALLISTIC MISSILE DEFENSE PROGRAMS

Antiship cruise missile defense, overland cruise missile defense, and ballistic missile defense (BMD) will all be necessary for naval (and joint) forces conducting 21st-century military operations for a number of reasons:

- ASCMD—Antiship cruise missiles in the hands of potential adversaries are numerous, sophisticated, and widespread. Every naval combatant becomes a target whenever it enters a theater and must defend itself effectively if it is to be an asset rather than a liability.

• OCMD—In the future, land attack cruise missiles will allow potential adversaries to deny military forces access to ports, airfields, and other entry points. In effect, the Navy has no OCMD capabilities, and building such capabilities will require time and investment.

• BMD—Tactical ballistic missiles are widespread weapons of terror and potential mass destruction. Naval forces need capabilities to provide ballistic missile defense to ports, airfields, and other entry points until assets arrive in-theater from other Services. In the future, longer-range ballistic missiles will become more prevalent and an adequate theater ballistic missile defense will require defense in depth.

With the exception of developing a robust capability for OCMD, there is little disagreement within the Office of the Chief of Naval Operations (OPNAV) and the Navy acquisition community concerning missile defense programs. Moreover, all Navy ballistic missile defense programs are matched to funding limitations or Ballistic Missile Defense Organization (BMDO)-imposed cost constraints and as a result have adopted evolutionary development programs that defer the development of necessary capabilities until far into the future.[3]

In the likely event that budget levels will not be sufficient to fund all cruise and ballistic missile defense efforts fully, the committee believes that the Department of the Navy will need to assign funding priorities for R&D efforts as follows:

1. ASCMD,
2. Area defense of forces and assets ashore against both overland cruise missiles and ballistic missiles (NAD system), and
3. The NTW system.

The committee's rationale for according first priority for R&D funding to ASCMD is that if the Navy does not have a robust ASCMD capability, its abilities to undertake or support operations in littoral areas will be seriously limited.

The committee could not come to a consensus on the relative prioritization of R&D funding between OCMD and NAD. All members of the committee

[3]The committee is also concerned that where naval R&D needs and priorities are not supported by BMDO investment, there is no safe mechanism for the Department of the Navy to apply funding of its own. Furthermore, the committee believes that if the Department of the Navy allocates R&D funds for theater missile defense, congressional committees will most likely cut those funds on the basis that missile defense R&D has already been accounted for in the BMDO budget. In the end, there is no investment for theater missile defense R&D. Therefore, the committee believes that a stronger organizational link should be established between the Department of the Navy, the Office of the Secretary of Defense, and the Ballistic Missile Defense Organization in order to support R&D.

recognized that defense against land attack cruise missiles and defense against ballistic missiles are necessary components of the same mission, particularly if the Navy is to protect forces and assets ashore.

Some committee members argued that since ballistic missiles are widely available to probable or potential adversaries and since land attack cruise missiles currently are not widely proliferated, priority for R&D funds should be assigned to the NAD program. Furthermore, ballistic missiles, which may be configured to carry weapons of mass destruction, can have a major political impact on allies and on forces ashore.

Others on the committee argued that the development of an OCMD capability (be it naval or joint) was essential for the protection of forces ashore against a threat that would be highly likely to proliferate if no such defense were to be developed. Those who supported a relatively high priority for R&D funding for OCMD also pointed out that the most effective means of developing an OCMD capability is through the use of an elevated detection platform. The same elevated platform and sensor system that is needed for OCMD can be used to extend the detection horizons of a surface ship. Thus, the sensor developments that will be necessary to provide OCMD capabilities will also help to improve the Navy's ASCMD capabilities.

Although the committee could not achieve a consensus on the relative priority for R&D funding between OCMD and NAD, it was very concerned that R&D funding for the development of a competent OCMD capability has been relatively limited. Unless R&D funding for OCMD is given higher priority than it currently has, the prognosis for the development of OCMD capabilities will continue to be bleak. Furthermore, if the Navy cannot provide OCMD in support of Marine or Army forces ashore, at least in the early stages of operations, then the full potential of naval expeditionary forces (as envisaged in Forward...From the Sea and OMFTS) will not be achieved.[4] Thus, without a land attack cruise missile defense capability to supplement their ballistic missile defense capabilities, naval forces' ability to influence events ashore will be limited to attacks on stationary targets with standoff missiles and air-delivered ordnance.

In its assessment of the Navy's existing and planned ballistic missile defense capability, the committee emphasizes the NAD system over the NTW system.[5] The basis for this emphasis on NAD relates to BMDO's role in de-

---

[4]Some might argue that in a developed theater the Army's Patriot advanced capability-3 (PAC-3) would be deployed. As currently configured, PAC-3 does not depend on the availability of an elevated air moving target indication (AMTI) radar to detect and track missiles that make maximum use of terrain obscuration in order to evade detection by ground-based radars. Thus, until PAC-3 is provided with a robust capability to negate missiles that employ terrain-obscured trajectories, no OCMD capability exists.

[5]Program Budget Decision 224 calls for a shift of $121 million from the NTW program to the NAD program over FY02 and FY03 (*Inside the Pentagon*, January 18, 2001, pp. 12-13).

fense-related development and acquisition for TMD systems. In some developed theaters competent land-based theater missile defense systems might be predeployed. For example, if the Army's theater high altitude air defense (THAAD) system were successfully developed and deployed, it could provide significant midcourse engagement capabilities in a theater where it had been deployed prior to the onset of conflict. In addition, if the Air Force's airborne laser (ABL) system were similarly successful, it could provide ascent-phase engagement capabilities against shorter-range ballistic missiles. In such circumstances, the NTW system would supplement the projected capabilities of these systems in addition to the projected endo-atmospheric ballistic missile engagement capabilities of both the NAD and Army PAC-3 systems.

**Recommendation: The Secretary of the Navy, the Chief of Naval Operations (CNO), and the Commandant of the Marine Corps (CMC) should assign R&D funding priority in the following order: (1) antiship cruise missile defense, (2) area defenses against both overland cruise missiles and ballistic missiles (NAD system) for the protection of forces and assets ashore, and (3) the NTW system.**

## ES.3 STOVE-PIPED THEATER MISSILE DEFENSE SYSTEMS[6]

The committee recognizes that the distributed architectures envisioned for future theater missile defense operations, driven by the realities of the availability and the readiness of defense elements, make it a risky and uncertain business to provide the required level of protection against threatening ballistic and cruise missiles. A significant part of the uncertainty associated with connecting available sensors and shooters into an effective defense network comes from the fact that TMD systems are developed and tested largely as vertically integrated defense systems (as, for instance, are NAD and PAC-3) and are relatively loosely integrated as a family of systems. This suggests that if dynamically assembled distributed architectures are to function effectively, a new paradigm for development and testing needs to be applied by BMDO and the Services.

**Recommendation: The Secretary of the Navy, the CNO, and the CMC should support the expansion of distributed defense development and test plans by the Ballistic Missile Defense Organization (BMDO) and experiments to demonstrate the related advanced engagement modes. To the extent practicable, the system integrated tests being planned by BMDO and experimental programs such as the Theater Missile Defense Critical Measure-**

[6]The term "stove-pipe" refers to a program that stands alone, i.e., is planned, constructed, and supported without regard to other programs within the Department of the Navy or within the other Services.

**ments Program should be structured and extended to incorporate the critical defense functions unique to distributed architectures.**

## ES.4 LIMITATIONS RELATED TO THE CONCEPT OF OPERATIONS FOR THE CONDUCT OF OCMD AND TBMD IN THE COURSE OF EXPEDITIONARY WARFARE OPERATIONS WHEN JOINT AND COALITION FORCES ARE PRESENT

The Navy has declared expeditionary warfare that will influence events ashore as one of its main missions. The Marine Corps Expeditionary Maneuver Warfare 21[7] (EMW 21) strategy is consistent with and dependent on the Navy's capability in this area. Expeditionary warfare and theater missile defense are thus mutually dependent.

Expeditionary operations envision the possibility of forcible seaborne entry into a theater in which Marine Corps forces are launched from Navy ships and proceed directly to targets beyond the shoreline. Such operations may include peace enforcement, noncombatant evacuations, or combat operations. For Marine Corps forces to have the required reach, it will be necessary that ship formations approach the shoreline as needed to deliver supporting fire and logistical support to the Marines ashore. The same kind of support could be required if Army elements are involved as part of a joint task force. In any scenario in the littorals, the Navy must be able to defend both its own ships and the assigned forces against attacks by ballistic and cruise missiles.

**Recommendation: To achieve a competent cruise missile defense capability for the support of naval and joint forces operating in littoral areas, the CNO and the CMC should do the following:**

- **Develop a concept of operations with the other Services that routinely substitutes and employs assets such as the airborne warning and control system (AWACS) air moving target indication (AMTI) radar or the joint land**

[7]Expeditionary Maneuver Warfare 21 is the Marine Corps overarching strategy for conducting 21st-century Marine Corps operations such as those described in "Operational Maneuver From the Sea"; "Ship to Objective Maneuver" (Van Riper, LtGen Paul K., USMC, 1997, "Ship to Objective Maneuver," Marine Corps Combat Development Command, Quantico, Va., July 25, available online at <http://192.156.75.102/stom.htm>); "Maritime Prepositioning Force 2010 and Beyond" (Krulak, Gen C.C., USMC, 1997, "Marine Prepositioning Force 2010 and Beyond," Headquarters, U.S. Marine Corps, Washington, D.C., December 30, available online at <http://192.156.75.102/mpf.htm>); "Sustained Operations Ashore" (Krulak, Gen C.C., USMC, 1998, "The Marine Air Ground Task Force in Sustained Operations Ashore," U.S. Marine Corps, Washington, D.C., June 10, available online at <http://192.156.75.102/soa.htm>); and "Other Expeditionary Operations" (Warfighting Requirements Division, to be published, "Other Expeditionary Operations, Draft Concept Paper," Marine Corps Combat Development Command, Quantico, Va.).

**attack cruise missile defense elevated netted sensors (JLENS) system to perform over-the-horizon target acquisition and missile command functions envisaged for the E-2C Radar Modernization Program (RMP) radar; and**

- **Leverage joint experimentation in order to develop the operational concepts and technical capabilities necessary for joint missile defense operations.**

## ES.5 ASCMD, OCMD, AND TBMD DEFICIENCIES AND THE PROGRAMS TO CORRECT THEM

Over the past several years, lower levels of R&D investment have allowed the ASCM threat to evolve somewhat more rapidly than shipboard defenses have been improved. Future threats, which are projected to have much smaller radar signatures, greater agility, and electronic countermeasure (ECM)-resistant sensors, may well overstress these defenses when the Navy is constrained to operate in a littoral environment. The proposed acquisition and deployment of SPY-3 and the X-band horizon search MFR, along with some advances in the Navy's electronic warfare techniques, should redress some but not all of the Navy's projected ASCMD deficiencies. The committee is concerned that there are no programs in place to develop additional techniques to increase the Navy's ASCMD effectiveness.

In the final analysis, the ASCMD problem relates to the fact that a low-altitude cruise missile can get relatively close to a surface ship before it crosses the radar horizon of the ship's defensive sensors. If the number of incoming cruise missiles is sufficiently large, their agility and speed sufficiently high, and their radar cross section sufficiently low, the defensive system will be overwhelmed. A strong layer of short-range self-defense is needed, but robust defense requires a depth of fire that can be provided only by employing elevated sensors, such as the JLENS, that extend the horizon of the defensive sensors, along with the use of a missile that is designed to intercept targets beyond the line-of-sight horizon of the firing platform. The committee was not briefed on any systems other than the Army's JLENS for solving this ASCMD problem.

With respect to OCMD, the committee observes that there is still no program that will provide a means for the ship-based defense of forces ashore against cruise missile attacks. Although ship-launched interceptor missiles of suitable range are available, the sensors that would permit them to engage cruise missiles not observable from the ship have not been developed or otherwise acquired. The Navy will have to develop the necessary airborne sensors to support an OCMD capability or seek ways in which systems of the other Services, such as JLENS, might be brought into position and employed.

**Recommendation: The Secretary of the Navy, the CNO, and the CMC should support the development of a competent cruise missile defense against anti-ship and overland cruise missiles. Beyond supporting the programmed devel-**

**opment and acquisition of multifunction radar (MFR) and volume search radar (VSR), such a capability should include the following components:**

- **An elevated AMTI radar—possibly AWACS or unmanned aerial vehicle (UAV)-based—with robust overland clutter negation capabilities and with future capabilities to operate in a multistatic mode so that low-radar-cross-section overland targets can be engaged;**
- **An overland, over-the-horizon variant of the SM-2 missile with dual-mode, semiactive, and active terminal guidance; and**
- **The extension of cooperative engagement capability (CEC) to allow the employment of air-directed surface-to-air missiles (ADSAMs) against targets that are beyond the line-of-sight horizon of weapon launch platforms.**

**Recommendation: Beyond supporting the SPY-1 upgrades to improve NAD and NTW discrimination capabilities, the Secretary of the Navy, the CNO, and the CMC should pursue an aggressive R&D effort aimed at producing the following capabilities:**

- **A high-resolution, X-band adjunct to the S-band SPY-1 radar that will allow discrimination among warheads, decoys, and debris and reduce the need for salvo launches;**
- **A hit-to-kill (HTK) vehicle with greater agility, divert capability, and lethal radius than the Block I HTK vehicle, giving it the ability to handle tethered and tumbling target complexes;**
- **A multicolor infrared sensor with improved sensitivity to extend acquisition ranges against low-infrared-signature targets and aid in discrimination; and**
- **A radar and/or LADAR on the hit-to-kill vehicle that could precisely measure body dynamics for effective discrimination against replica decoys.**

**Recommendation: In an effort to examine countermeasures beyond the design threat of naval theater ballistic missile defense systems, the Department of the Navy should maintain an ongoing red-blue effort that provides continuous analysis, design, and testing of potential theater ballistic missile defense countermeasures and defense responses and works closely with corresponding Ballistic Missile Defense Organization (BMDO) efforts. This effort could be conducted in a manner similar to the prior Advanced Ballistic Reentry System Program, which developed penetration aids for U.S. intercontinental ballistic missile systems, or an extension of the current project Hercules, supported by BMDO, that is looking at advanced discrimination techniques.**

## ES.6 CURRENT AND PROJECTED MARINE CORPS OCMD CAPABILITIES

Marine Corps plans for OMFTS and Ship-to-Objective Maneuver (STOM) depend on shipboard basing of assault elements and rapid transport of light forces to inland objectives. The Navy is expected to provide air support—close air support along with Marine Corps air, combat air patrol, ship-based fire support, and ship-based early warning of and defense against air and ballistic missile attack. The Marine Corps is also dependent on the Navy for logistical support of many kinds. In the future, the Corps will have a ground-launched advanced medium-range air-to-air missile capability—a complementary low altitude weapon system—light enough to be taken ashore with assault units but with limited sensor capability, necessitating CEC cueing.

**Recommendation: Recognizing that there will always be some gaps in naval air defense coverage due to extended littoral operations, the Secretary of the Navy, the CNO, and the CMC should support the development and acquisition of the complementary low altitude weapon system (CLAWS) and the multirole radar system (MRRS); interfaces should be developed for targeting and fire control to the following sensors:**

- **Army JLENS radar system,**
- **Marine Corps TPS-59 (V-3) radar system,**
- **E-2C RMP AMTI radar, and**
- **Air Force AWACS SPY-1/2 radar system.**

**Recommendation: Recognizing that the MRRS may not be ready in time to provide an initial targeting and fire control radar for the CLAWS, the Secretary of the Navy, the CNO, and the CMC should consider deployment of the TPS-59 radar on designated maritime preposition force squadrons as an interim measure.**

## ES.7 BATTLE MANAGEMENT COMMAND, CONTROL, AND COMMUNICATIONS (BMC3)

A commander must have the means to understand the operational environment, the location and condition of his forces, and the actions of the enemy. He/she must be able to communicate well enough to reallocate resources and vary subordinate assignments as appropriate to achieve a particular mission, keeping superiors advised as necessary. The need today to comprehend and control on a theaterwide basis presents an immense challenge.

**Recommendation: Given that management of battle-space force components is a critical aspect of missile defense that is currently seriously defi-**

**cient, Department of the Navy leadership should actively support efforts relating to doctrine, acquisition programs, and research to overcome such deficiencies, in particular by:**

- **Supporting current efforts such as the Single Integrated Air Picture (SIAP) System Engineering Office Program, which is seeking to enhance the quality of the air-space picture;**
- **Supporting the development of concepts of operations necessary for expeditionary and joint Service littoral operations, including means for offense-defense coordination;**
- **Recognizing that for success in these operations the Department of the Navy will require support from other Services; and**
- **Recognizing that all battle-space management development efforts must seek to accommodate the inclusion of unplanned force components.**

**Recommendation: Given that Link 16 and CEC, even when evolved and improved, will not provide a full battle management command, control, and communications (BMC3) capability for either overland cruise missile defense or theater missile defense, the Department of the Navy leadership should initiate actions leading to the development of a next-generation BMC3 system. This entirely new system, leveraging both commercial and defense technology advances, should include the following features:**

- **Support of highly flexible and adaptable combinations of naval and joint force configurations by allowing assets to interface readily with one another (e.g., through an Internet Protocol-based, quality-of-service-guaranteed infrastructure);**
- **Wide-bandwidth, bandwidth-on-demand wireless communication networks with dynamic allocation of resources; and**
- **Initial development of a prototype in parallel with existing BMC3 systems to encourage experimentation and adoption.**

**In addition, development of a high-bandwidth test bed would be particularly valuable. It would allow new capabilities to be tested and explored in the near term while the existing BMC3 systems continue to undergo their intended evolution; transition to the new capabilities would occur only after they had been adequately developed and accepted.**

## ES.8 TECHNOLOGY INVESTMENT

As presented to the committee by the Navy and Marine Corps, the developmental paths intended to evolve TMD capabilities are generally reasonable, although several exceptions are identified in this report. The evolutionary, or "spiral," development of added capabilities to pace the threat is a reasonable concept. However, the committee is concerned that the technology required to support the intended evolution is not being developed. The necessary investments must be made to bring the required technology to a state where it is available for use in the time frame intended.

**Recommendation: In its technology investment program, the Department of the Navy should develop sensors, weapons, and BMC3 architectures and algorithms that are adaptive and flexible enough to allow responding to unexpected threat capabilities and characteristics. These ballistic missile defense system elements should be combined into experimental systems for evaluation and refinement. The mature technologies from the program should be incorporated into future spirals of the NAD and NTW ballistic missile defense systems.**

# 1

# Introduction and Background

Cruise missiles and ballistic missiles exist in large numbers in the tactical inventories of many potentially hostile nations. These weapons vary in their ability to evade detection by the military forces' defensive systems and in the accuracy with which they deliver warheads. In any case, however, they have a demonstrated ability to sink or seriously damage ships and to deliver warheads with high precision against stationary land targets. Among the more troubling aspects of the missile threat is the fact that almost all such weapons may be modified to transport nuclear, chemical, or biological warheads.

Since the development of guided missile systems in the 1950s, the Navy has had vigorous programs to protect itself against air threats. Shipboard antiair warfare (AAW) systems have been deployed for both ship self-defense and for area defense. The Navy developed air-to-air weapon systems to protect its tactical aircraft and to serve as an outer layer of ship defense. The AAW systems deployed by the Marine Corps have been primarily ground-based air defense systems. The relative importance of each of these systems has waxed and waned as naval missions, operational concepts, and threats evolved.

In this report, current and planned naval capabilities for theater missile defense are divided into three broad categories:

- Antiship cruise missile defense (ASCMD),
- Overland cruise missile defense (OCMD), and
- Theater ballistic missile defense (TBMD).

The Navy's TBMD capabilities are further subdivided into the Navy area defense (NAD) and Navy theater wide (NTW) systems. If successful, NAD systems are designed to engage reentry vehicles (RVs) delivered by ballistic missiles after they have reentered the atmosphere. NTW systems are designed to engage threats at exo-atmospheric altitudes. This subdivision reflects the fact that the Navy is planning to use different weapon systems for endo-atmospheric and exo-atmospheric engagements.

The issues mentioned in this introduction are discussed in more detail in Chapters 2, 3, and 4. Chapter 5 presents the committee's conclusions and recommendations.

## 1.1 CRUISE AND BALLISTIC MISSILE THREATS

Through their evolving strategies Forward...From the Sea[1] and Operational Maneuver From the Sea (OMFTS),[2] the Navy and Marine Corps have acknowledged a shift in warfare from operations on the open seas to operations in and adjacent to littoral areas. This shift in warfare location presents many technical and operational challenges to naval forces in power projection, the most notable of which may be an increase in the land-based threat to the forces engaged in such operations.

Both TBMD and cruise missile defense (CMD)—including OCMD and ASCMD—are important emerging military capabilities that will be needed if naval forces are to execute missions in and near littoral areas. Today, there are large numbers and varieties of cruise and ballistic missiles in the operational inventories of many potential future adversaries of the United States.

Although high-performance ballistic missiles exist and could become available to potential future adversaries, most of the ballistic missiles that are currently available to such adversaries are of rather unsophisticated design. Many have limited accuracy of delivery and are ineffective for hitting tactical targets. As currently configured, many are nonseparating, single-stage rockets that are less stressing to defense systems than multistage missiles. Many others are not able to deploy penetration aids. In a military sense, these threats will have limited tactical value unless they carry nuclear, chemical, and/or biological warheads. However, even as currently configured, they pose a serious threat to deployed forces and assets, as well as to the political stability of neighboring or allied countries.

---

[1]Department of the Navy. 1994. "Forward...From the Sea, Continuing the Preparation of the Naval Services for the 21st Century," U.S. Government Printing Office, Washington, D.C., September 19.

[2]Headquarters, U.S. Marine Corps. 1996. "Operational Maneuver From the Sea," U.S. Government Printing Office, Washington, D.C., January 4. Available online at <http://192.156.75.102/omfts.htm>.

As was evident from the information presented to the committee, there are many reasons to expect that in the near future (5 to 20 years) more sophisticated variants of these ballistic missiles will proliferate. The committee observed that a number of design improvements to ballistic missiles had shown up earlier than expected, portending a significant increase in the effectiveness as well as the number of missile threats that will be faced. It anticipates that these variants will be characterized by an ability to be employed against tactical targets, an ability to deploy sophisticated penetration aids to counter defensive systems, and an ability to maneuver to evade defensive interceptors. Thus, any TBMD system that the Department of the Navy develops and fields must be considered in the context of such plausible future threats.

The sophistication of cruise missiles that are currently available to potential adversaries can also stress our military forces' defensive systems seriously. Cruise missiles that can be purchased from France, Russia, and China (or indigenously manufactured) may be characterized by their low-altitude trajectories, their high velocity, their low nose-on radar cross section (RCS) values, and sensors that are robust against many forms of electronic countermeasures (ECMs). Alternatively, adversaries may elect to attack our military forces in the littorals with a large number of relatively inexpensive, low-technology cruise missiles in an attempt to overwhelm our defenses by the sheer numbers.

Given these factors, the committee believes that the antiship cruise missile (ASCM) threat will intensify, especially as naval operations shift to the littorals. Defense against current state-of-the-art antiship cruise missiles (ASCMs) and overland cruise missiles (OCMs) represents a serious challenge for both existing shipboard and expeditionary warfare defensive systems and the new generation of systems now under development.

## 1.2 EXISTING THEATER MISSILE DEFENSE CAPABILITIES

### 1.2.1 Antiship Cruise Missile Defense Issues

In evaluating the effectiveness of an ASCMD system, the threat cruise missile's velocity, RCS, and altitude above sea level must be considered. When these parameters are combined in a three-dimensional plot, it can be seen that there are some regions in this parameter space where the incoming missile could not be shot at and other regions where only one or two shots would be possible. The combinations of missile velocity, RCS, and altitude where current defensive systems can exercise a shoot-look-shoot (SLS) doctrine are limited. Some current threat cruise missiles are designed to operate in the regime where only one defensive interceptor launch is possible. In certain circumstances, when high-Mach-number, low-RCS, and low-trajectory cruise missiles are fired from a high-clutter land mass, it may not be feasible to launch any current defensive interceptors.

Unless the Navy's currently deployed sensor systems, interceptors, and weapon control systems are improved, postulated and technically realizable ASCM improvements will enable ASCMs to operate in speed, altitude, and RCS regimes where they cannot be engaged. The committee believes that the Navy has a number of important developments under way to improve its ASCMD capabilities. These include the multifunction radar (MFR), which will provide a much better X-band horizon search capability, the ship self-defense system (SSDS), which will provide non-Aegis ships with rapid response weapon control, and the evolved sea sparrow missile (ESSM), which will greatly improve engagement with advanced threats. These capabilities need to be deployed as soon as possible.

### 1.2.2 Overland Cruise Missile Defense Issues

The current U.S. Marine Corps operational strategy, Expeditionary Maneuver Warfare 21 (EMW 21),[3] envisages the use of light and highly mobile forces that are largely unencumbered by large, heavy air defense and artillery systems. The Marine Corps is looking to forces at sea to provide OCMD and fire support. Current Marine Corps concepts of operations for OCMD are based on an elevated air moving target indication (AMTI) radar with overland clutter rejection capabilities as an essential feature. The radar system that is currently available in the Navy's E-2C aircraft is inadequate for such purposes. The E-2C Radar Modernization Program (RMP) was intended to develop and deploy a suitable AMTI radar. If deployed, this radar would be the first Navy AMTI radar with significant overland detection capability. The E-2C RMP, in conjunction with the cooperative engagement capability (CEC) data link and a postulated dual-mode active/semiactive variant of the standard missile (SM)-2, would provide the Department of the Navy with an expensive (and possibly vulnerable) but respectable OCMD capability.

Funding constraints are currently jeopardizing the RMP and casting doubt on the achievement of a credible OCMD capability. If funding limitations preclude the development of a sea-based AMTI capability, alternatives must be

[3]Expeditionary Maneuver Warfare 21 is the Marine Corps' overarching strategy for conducting 21st century Marine Corps operations such as those described in "Operational Maneuver From the Sea"; "Ship to Objective Maneuver" (Van Riper, LtGen Paul K., USMC, 1997, "Ship to Objective Maneuver (STOM)," Marine Corps Combat Development Command, Quantico, Va., July 25, available online at <http://192.156.75.102/stom.htm>); "Maritime Prepositioning Force 2010 and Beyond" (Krulak, Gen C.C., USMC, 1997, "Marine Prepositioning Force 2010 and Beyond," Headquarters, U.S. Marine Corps, Washington, D.C., December 30, available online at <http://192.156.75.102/mpf.htm>); "Sustained Operations Ashore" (Krulak, Gen C.C., USMC, 1998, "The Marine Air Ground Task Force in Sustained Operations Ashore," U.S. Marine Corps, Washington, D.C., June 10, available online at <http://192.156.75.102/soa.htm>); and "Other Expeditionary Operations" (Warfighting Requirements Division, to be published, "Other Expeditionary Operations, Draft Concept Paper," Marine Corps Combat Development Command, Quantico, Va.).

explored. The committee believes that the Navy and Marine Corps should explore alternative concepts of operations such as the coordinated use of the Air Force airborne warning and control system (AWACS) aircraft, use of the Army's joint land attack cruise missile defense elevated netted sensors (JLENS) system, the deployment of a multiple unmanned aerial vehicle (UAV) surveillance system, and/or the use of National sensors.[4]

Even if the RMP funding problems are resolved, projected progress in cruise missile capabilities indicates that the Navy will eventually need to provide performance beyond that of the RMP AMTI radar operating monostatically with the ADS-18 antenna. A network of distributed, multistatic sensors could provide a reliable capability to negate low-observable cruise missiles. The committee believes that the Navy should put more emphasis on advanced R&D in support of improvements to the Department of the Navy's OCMD capabilities.

In keeping with its commitments to the concepts of OMFTS and Ship-to-Objective Maneuver (STOM),[5] the Marine Corps has already decommissioned its Hawk[6] capabilities. It has embarked on a program to develop a lightweight, mobile missile defense system utilizing an advanced medium-range air-to-air missile (AMRAAM) mounted on a high-mobility, multipurpose wheeled vehicle (HMMWV). The resulting system, called the complementary low altitude weapon system (CLAWS), should provide the Marine Corps with a limited forward-deployed air defense capability. The major problem with CLAWS is that the Marine Corps TPS-59 (V3) radar may be the only sensor (in the near term) available to provide it with target information.

Although the TPS-59 (V3) is an excellent long-range air surveillance radar, it is large and difficult to transport in current amphibious shipping. Under current concepts of operations (CONOPS), the TPS-59 (V3) is unlikely to be available in the execution of OMFTS or STOM missions to support forces deployed ashore or Navy ships operating offshore. The committee believes that in such situations, the Marine Corps will need to depend on either an elevated radar with good overland performance (e.g., the E-2C RMP AMTI radar with the ADS-18 antenna) or a compact, high-performance, mobile, ground-based air surveillance radar. The committee further believes that in the years before a high-performance elevated AMTI radar—AWACS or a sea-based system—is routinely available for forward-deployed forces, the Marine Corps should rethink its concept of operations and consider placing the TPS-59 in its maritime preposition force (MPF) squadrons for a rapid response capability in support of OMFTS or STOM missions.

---

[4]The term "National" refers to those systems, resources, and assets controlled by the U.S. government but not limited to the Department of Defense (DOD).

[5]Van Riper, LtGen Paul K., USMC. 1997. "Ship-to-Objective Maneuver," Marine Corps Combat Development Command, Quantico, Va., July 25. Available online at <http://192.156.102/stom.htm>.

[6]Originally named for the predatory bird, but later the name was turned into an acronym for "homing all the way killer."

The Office of Naval Research (ONR), the Marine Corps, and the Army have expressed interest in the development of an air surveillance radar system called the multirole radar system (MRRS) mounted on an HMMWV. At the time this study was carried out, the MRRS was not a funded program. The committee believes that if MRRS can perform in accordance with its postulated nominal specifications, it would be suitable for the operations in which the Marine Corps envisions becoming involved.

### 1.2.3 Navy Area Defense and Navy Theater Wide Issues

In recognition of the importance of ballistic missile defense (BMD), the Navy is developing the NAD and the NTW systems. The NAD system will employ the SM-2 Block IVA, and the NTW system will employ the SM-3. Both missiles are derivatives of the mature SM-2.

Upgrades appear warranted to extend the sensitivity, discrimination capabilities, and target acquisition range of the infrared (IR) seekers proposed for the SM-2 Block IVA and the SM-3, especially to contend with future threats. Currently, the limited target acquisition range of the interceptor missile's seeker results in a very compressed end-game time line that might cause the performance of these weapons to be degraded by signature-reducing IR countermeasures such as warhead coatings or shrouds. Such countermeasures would further reduce the target acquisition range of the seekers.

As currently programmed, all indications are that the NAD system is progressing through its test program successfully. The committee found the NAD system to be reasonably structured and to have manageable risks. However, it was concerned that the current system is not supported by a robust R&D program that will provide for preplanned product improvement (P3I) to allow for matching the upgrades of defensive capabilities to the upgrades of future threat ballistic missile capabilities.

The committee notes that the NTW system is not as mature as the NAD system. The Achilles' heel of this program is the SPY-1 radar, which is an excellent air defense radar but a marginal radar for the full range of NTW mission requirements. As currently configured, it is not adequate to support the capabilities required of a theater-wide ballistic missile defensive interceptor. For ascent-phase engagements, which may be an important function for NTW, the large RCS of the target booster may support adequate fly-out of the interceptor; however, this engagement mode cannot be counted upon for all scenarios. In many situations, the SPY-1 will require external cueing by overhead sensors such as the space-based infrared sensor-low (SBIRS-low). Unfortunately, SBIRS-low is not progressing at a pace that can inspire confidence that it will be available in time to assume the cueing function for the NTW system.

Future ballistic missiles that the NTW is likely to be called upon to negate may be sophisticated enough to launch decoys and penetration aides and to maneuver. A system that is designed to perform in a robust manner against such missiles should have capabilities that are considerably more advanced than those that will be available with the Block-I NTW system as it is presently designed.

Ideally, the exo-atmospheric hit-to-kill (HTK) warhead should have a multicolor IR sensor to help distinguish an incoming warhead from decoys and debris. In addition, radar that distinguishes a spinning body from a nonspinning body would be an important capability, as would laser identification and ranging (LIDAR) that permitted incoming warheads, decoys, and debris fields to be imaged. Certainly there would be other important capabilities for an HTK vehicle: greater agility, divert capability, and lethal radius that would allow it to engage tumbling but connected warheads and decoys. Should the HTK strategy prove inadequate, some thought has been given to a "kill enhancement" device to be added to the kill vehicle, but no alternatives for improved lethality are funded at present.

Perhaps the greatest shortcoming of the Block-I NTW system is the lack of a high-resolution X-band adjunct to the S-band SPY-1 radar. Ultimately, a radar will be needed that will allow discrimination among warheads, decoys, and debris at ranges that support earlier interceptor engagement. Without such a discrimination capability, the defending system is required to fire a salvo at the ensemble of incoming targets rather than only at warheads. Against the small RV cross sections of the more advanced threats that exist today, there may not be any shot opportunities. The committee is concerned that no R&D programs are planned to address these issues. The plan for the Block-II phase of NTW does not appear to be completely developed. The advanced technology programs in support of NTW do not appear to be tied to requirements for improved performance or meeting advanced threats. As an example, the information presented to the committee concerning NTW technologies that are under development cited "advanced seekers" but did not say what sensitivity or resolution would be achieved or which advanced threats the improved seekers could meet.

The committee recognizes that in the area of R&D to support new BMD capabilities, the Navy cannot proceed in a completely autonomous manner. R&D programs must be coordinated with and supported by the Ballistic Missile Defense Organization (BMDO). Since the Gulf War, BMDO programs have emphasized the acquisition of theater missile defense (TMD) systems and National missile defense rather than R&D. BMDO advanced technology programs have decreased in numbers and funding. As a result, the systems that are now being acquired do not have credible plans or resources for future block upgrades to correct deficiencies or meet advanced threats.

Most of the technologies needed to achieve baseline capabilities against the near-term TBM threat appear to be in hand or nearly so. However, there is a dangerous lack of the technology development programs that are necessary if the Navy is to achieve the defensive capabilities that will negate plausible future threats. Forward...From the Sea and OMFTS will not be viable strategies if the

Department of the Navy is unable to adequately protect itself (and joint forces) against the ballistic and cruise missiles threat projected for that era.

BMDO is coordinating the advanced technology program for the TMD systems programs that are being executed by the Services. Although there is certainly room for improvement in the execution of this program—for example, the committee believes that better traceability between postulated future threats and the technological responses to those threats would make better use of available resources—as previously noted, the fundamental problem faced here is inadequate resources.

The committee believes that the NTW program has an extremely aggressive schedule that postpones the hard problems to future upgrades. Unfortunately, there is no attendant R&D program dedicated to the solution of these problems. As an example, the earliest versions of the NTW will have inadequate detection capabilities; moreover, no development program is in place that will eventually alleviate these inadequacies. As currently funded, the program is heavily reliant on congressional appropriations that go beyond Service requests.

### 1.2.4 Concepts of Operations in Combined Offensive-Defensive Environments

Central to the effective utilization of these technologies are concepts for executing the missile defense missions in an operational theater. The committee sought, during several of the Navy and Marine Corps briefings, to understand the concept of operations that would be used for expeditionary operations. In particular, the committee wished to learn how the theater missile defense operations might be coordinated with other operations that take place at the same time and in the same area. Various presentations indicated that aircraft would be delivering ordnance and providing logistical support to Marine Corps units ashore. Simultaneously, ships would execute fire missions by launching extended-range guided missiles (ERGMs) and other land attack weapons as called for by the Marine Corps. The committee believes that it is necessary to formulate a concept of operations that does everything to ensure that theater missile defense can be undertaken in coordination with offensive operations. Both offensive and defensive activities must succeed without confliction or danger to friendly forces.

Joint doctrine[7] has been promulgated to guide the conduct of air and missile defense in a theater; however, the doctrine appears to presume that the theater

[7]Fulford, LtGen C.W., Jr., USMC, Director, Joint Staff. 1999. "Joint Doctrine for Countering Air and Missile Threats," Joint Publication 3-01, The Pentagon, Washington, D.C., October 19. Available online at <http://www.dtic.mil/doctrine/jel/new_pubs/jp3_01.pdf>; Ross, Lt Gen Walter K., USAF, Director, Joint Staff. 1996. "Doctrine for Joint Theater Missile Defense," Joint Publication 3-01.5, The Pentagon, Washington, D.C., February 22. Available online at <http://www.dtic.mil/doctrine/jel/new_pubs/jp3_01_5.pdf>.

has already been developed and that joint forces are in place. Other than to note that the same functions must be performed in undeveloped theaters, the joint doctrine does not give guidance for expeditionary warfare. Although the doctrine is being evolved at JTAMDO[8] and "early entry" operations are mentioned in the draft 2010 Operational Concept, the doctrine for offense/defense coordination is developed only to the point where the two functions are said to be synchronized.

Pending the development of joint doctrine to guide initial operations in an undeveloped theater, it falls to the Navy and Marine Corps to develop the appropriate CONOPS. A CONOPS for expeditionary warfare in the littorals must address conflicting requirements for the employment of operational assets and the control of offensive and defensive operations.

Concepts for conduct of the offense are amenable to preplanning to avoid conflict yet must remain flexible enough to support operations ashore by Marine Corps units whose plans may have to be changed because of real-time events. At the same time, and in the same area, defensive measures must be taken to defeat ballistic missile, cruise missile, and aircraft threats to forces afloat and ashore. The conduct of effective theater missile defense without disrupting or conflicting with offensive measures is very difficult but necessary.

Several briefers[9] informed the committee that concepts for coordinating offensive and defensive operations have not been worked out. Such coordination concepts deserve considerable effort, since they are critical to the conduct of expeditionary warfare and necessary for evaluating the adequacy of theater missile defense programs.

### 1.2.5 Battle Management Command, Control, and Communications (BMC3) Issues

BMC3 must support joint as well as purely naval operations. Underlying BMC3 is the exchange of information among the missile defense sensor, command, and weapon nodes. The primary vehicle planned for this is the Link 16 networking and messaging scheme, as realized in the joint tactical information distribution system (JTIDS) radio terminals. These terminals will be installed on a variety of aircraft, surface ships, and submarines over the next several years, as well as in Patriot and theater high altitude air defense (THAAD) forces. JTIDS was first developed at least 30 years ago; thus, even though JTIDS is just now being deployed, it is very much a legacy capability.

---

[8]Joint Theater Air and Missile Defense Organization and Ballistic Missile Defense Organization. To be published. "Annex G: GTAMD 2010 Operational Concept (Draft Version 5 (Unclassified—For Official Use Only))," from *1999 Theater Air and Missile Defense (TAMD) Master Plan (U),* The Pentagon, Washington, D.C., December 2 (Classified).

[9]A listing of presentations to the committee is provided in Appendix F of this report.

The legacy nature of Link 16 is manifested by its lack of operational flexibility and its limited bandwidth. That is, the time division multiple-access networking scheme used by Link 16 is very complex to arrange and taxes operator skills. It can take a week or two to develop and test the scheme used in actual operation. Thus, Link 16 does not currently support flexible, rapidly conceived operations.

The current maximum bandwidth of the JTIDS radio in antijam mode is 115 kbps, and it is often much less in practice—very low figures by today's information transfer standards. A more modern approach based on commercial technology would allow much greater bandwidth. Commercial wireless technology is advancing rapidly, and capacities of at least a few megabits per second are currently possible. The commercial technology appears to have the necessary quality of service for military applications, although jam resistance is not a significant factor in commercial developments. Still, the commercial technology would offer a good base upon which to build a jam-resistant capability.

Several improvements are currently planned for Link 16. The joint interface control officer (JICO) position has been established to facilitate the difficult network management involved in Link 16 and the other tactical data links. Furthermore, the Space and Naval Warfare Systems Command (SPAWAR) is working to enhance Link 16 flexibility and bandwidth, among other things. These improvements have merit, and the committee supports them. However, they are best viewed as late-life upgrades to a system that is nearing the end of its technical life cycle.

Link 16 passes radar track (and other) data, as do all tactical data links. The CEC, by contrast, passes, combines, and produces measurement-level data from multiple radars and other available sensors to form a composite track picture in near-real time. Low-rate initial production of CEC components began in 1998, and operational evaluation is planned for the spring of 2001. The Navy's intent is to deploy CEC widely—on cruisers and carriers and on some destroyers, amphibious ships, and surveillance aircraft—although funding difficulties may limit the realization of this objective.

The composite track picture provides each CEC participant with a better track picture than that which the participant could generate alone, including higher-quality tracks and greater geographic coverage. This greater coverage will allow a given participant to launch its defensive missiles before its radar acquires a target—the so-called engage-on-remote and forward-pass concepts. Control of the defensive missiles being fired using CEC data lies in the command and decision module, which uses CEC data to control the firing of the defensive missiles. Thus, CEC does not form a complete BMC3 system nor was it intended to; loosely speaking, it is a "distributed sensor system."

The initial CEC focus is on battle group air (including cruise missile) defense. Additional uses are being considered that raise the question of whether they need extensive CEC capabilities or whether lesser (and presumably less expensive) capabilities would suffice. First, CEC is being planned for use in naval and joint ballistic missile defense. However, a ballistic missile track picture is much easier to obtain than a low-altitude cruise missile picture. Second, CEC is being proposed where there is little overlap in coverage between the sensors involved—for example, coupling Aegis and Patriot in low-altitude cruise missile defense. The main benefit of CEC appears to be its ability to provide a composite track picture from overlapping sensor coverage. In both cases the question is whether exchange of track data over a (possibly modernized) tactical data link would be adequate.

The committee believes a CEC capability would be valuable for battle group air defense but notes that the briefings it received did not provide analytical justification for using CEC instead of the presumably less expensive tactical data links in the cases just described. The committee cautions against use of CEC just because it is there, without adequate analysis of alternatives. Moreover, one should also be cautious about locking into CEC technology, and it must be kept in mind that it is based on an architecture first designed in the 1980s.

One of the most central BMC3 capabilities—and one of the most challenging to obtain—is a complete and accurate air space picture. Such a picture is necessary, for example, to use defensive assets efficiently and to coordinate the operation of offensive and defensive forces. In realistic situations this capability is achieved today in only a small fraction of the air space. There are a number of reasons for this, both procedural and technical, including lack of a common time reference, lack of navigation capability and its integration with the tactical network, connectivity shortfalls, failure to achieve a common geodetic coordinate frame, and differences in correlation/decorrelation algorithms. In part, the problems relate to the Link-16 networking and messaging scheme, but they are much broader than that. CEC addresses some of the problems, but again they are much broader than that.

To confront these problems, the Joint Requirements Oversight Council (JROC) directed in March 2000 that the Single Integrated Air Picture (SIAP) System Engineering Office be formed. The SIAP system engineer is responsible for the systems engineering necessary to develop recommendations for systems and system components that collectively enable building and maintaining a SIAP capability. Thus far, the office has identified candidate solutions to address problems such as those noted above. The near-term emphasis will be on engineering and recommending improvements to fielded systems, particularly the tactical data links.

The SIAP System Engineering Office's activities thus far appear well directed. The committee believes the Navy and Marine Corps should strongly support the activities of the office and monitor those activities to make sure they are meeting naval needs. The committee further believes that the SIAP system engineer should aggressively promote the development of modern alternatives that would eventually replace the current tactical data links.

Obviously the Department of the Navy and the Department of Defense (DOD) cannot abandon all legacy equipment in the near term to achieve a more flexible BMC3 capability. But if they do not starting thinking soon about what that improved capability will be, they will continue to be bound to the current legacy capability.

# 2

# Present and Projected Theater Missile Threats

## 2.1 TACTICAL MISSILE PROLIFERATION

The current proliferation of cruise missile technology is extensive and accelerating. A partial listing of the ASCMs that are being produced and sold on the worldwide market includes variants of the following:

- Aerospatiale Exocet,
- BAE Sea Eagle,
- IAI Gabriel,
- OTO Melara (Breda) Otomat,
- Saab RBS-15,
- MDAC (Boeing) Harpoon,
- SS-N-25 Harpoonski, and
- Russian SS-N-22, 26, and 27 supersonic missiles.

Briefings to the committee about the threat dramatically illustrated the scope of this proliferation. Several factors appear to be fueling this growth. The Gulf War and other conflicts made clear the political impact and value of such weapons to lesser powers and helped create a ready market in consumer countries whose wealth comes from oil.

Russia's need for hard currency has made it willing to market its most modern weapons. China's growing missile technology capability and apparent willingness to export that expertise, along with the marketing efforts of European weapon suppliers, make it likely that the United States will encounter significant numbers of these weapons in any future operations.

## 2.2 CRUISE MISSILE THREATS

The weapons referred to in this study under the generic title of cruise missiles can be ground- or air-launched. Ground-launched cruise missiles are generally multistage missiles, with the first stage being a rocket. At some predetermined altitude, the rocket booster is discarded and wings or canards are deployed to provide aerodynamic lift. Simultaneously, a motor is activated to provide propulsion. Air-launched cruise missiles are carried to launch altitude by an aircraft. In this case, too, tail fins, wings, and/or canards may be deployed to provide trajectory control surfaces.

A cruise missile may be accelerated to cruise speed by a rocket booster and might be designed to employ rocket thrust for a high-speed terminal attack. For most of its flight, a cruise missile is propelled by air-breathing turbojet or ramjet engines and relies on aerodynamic lift to carry its weight and maintain altitude. Cruise missiles remain within the atmosphere and under power during their cruise phase. Hence, their range and general flight characteristics are similar to those of an airplane.

Payloads carried by cruise missiles may include large, unitary, high-explosive warheads, submunitions, runway penetrators, or weapons of mass destruction (nuclear, chemical, or biological). In the past, successful cruise missile attacks succeeded in sinking warships or causing severe, mission-limiting damage to them. Cruise missiles that are configured to carry and dispense submunitions constitute a particularly severe threat to troops in the field and to nonarmored vehicles such as trucks. When the submunitions that are dispensed by a cruise missile are high-performance, self-propelled devices that are equipped with terminal engagement sensors, they can even constitute a significant threat to armored vehicles. Thus, cruise missiles are a significant threat both to platforms at sea and to forces ashore.

Cruise missiles can be classified according to the altitude and velocity of their cruise segment, as well as their launch-to-target range. Cruise altitudes fall into three categories: high altitude, low altitude, and surface skimming. High-altitude cruise extends the range by improving fuel-use efficiency, but it makes the cruise missile more likely to be detected. At lower altitudes, such missiles can take advantage of the decreased line-of-sight horizon for trackers in the vicinity of the target and of terrain features that mask the approach path. Surface skimmers, which are practical only over the ocean or extremely flat and desolate terrain, descend to within a few meters of the surface and may go undetected until very close to the target. Cruise missiles may cruise at subsonic, supersonic, or hypersonic velocity. Because lift-to-drag ratios decrease with increasing speed above the maximum endurance speed, a range penalty is paid for supersonic or hypersonic flight. However, since the time of flight is inversely proportional to speed, the intercept problem becomes more difficult as the speed of the target missile is increased. Of course, a cruise missile's flight path may be broken into segments with different altitude-speed characteristics to maximize the probability of mission success.

A cruise missile is not constrained to follow a single path to its target and can, in fact, follow a devious route to avoid obstacles and terrain, to hide below the tracker's line of sight, and to deceive defenders. While the thrust-to-weight ratio need not be large to maintain cruising flight, a cruise missile is easily designed to pull significant *g* load factors, allowing it to change direction quickly. Hence, it can jink, S-turn, and feint on the way to its target. It can approach its target a few meters above the surface or pull up and dive on its target at a high angle. To limit the ability of defending forces to maintain its trajectory in track, a given cruise missile can be programmed to choose apparently random approach maneuvers.

Unlike a ballistic missile engagement, a successful intercept of a cruise missile before it approaches its target virtually assures that the cruise missile will fail to accomplish its mission. Furthermore, less damage may be necessary to defeat a cruise missile than to defeat a ballistic missile. Such a missile need not always be totally destroyed—degraded performance in the form of diminished accuracy for the guidance sensors or a partial loss of aerodynamic control authority may be enough to cause it to miss its intended target.

Cruise missiles can attack both stationary and mobile targets. If a movable target is stationary for an extended period of time, the missile may be programmed to fly to the global positioning system (GPS) coordinate where the target is known to be located. Worldwide open access to the GPS and GLONASS (the Russian equivalent of GPS) networks simplifies the navigation and guidance systems for cruise missiles designed to attack fixed targets. Ten-meter navigational accuracy to any latitude-longitude pair is readily obtainable now that GPS selective availability (SA) has been turned off. One-hundred-meter accuracy is available with SA operating. In a major conflict, the United States might take measures to restrict the local availability of GPS to its adversaries. However, there is no precedent to indicate that during low-intensity operations, the operation of GPS will be restricted.

Cruise missile attacks on moving targets are multistep processes. First, the missile must be guided to fly to a point where, based on a target track developed by an external sensor, there is reason to expect that the terminal sensor on the cruise missile will be able to acquire the intended target. If the intended target is within the acquisition basket of the missile's seeker, the seeker can acquire the target and the missile can guide itself on a collision course to the target. A wide variety of seekers have been developed to support the terminal engagement phase of cruise missile attacks. If the cruise missile does not have a data link back to an individual who can evaluate the output of the cruise missile's sensor and control the terminal engagement, it must be guided to the target autonomously.

Autonomous guidance sensors are subject to jamming, deception, and decoys. ECMs against missile guidance and navigation systems have been employed since World War II, as have electronic counter-countermeasure (ECCM) techniques that are designed to negate the effects of defensive countermeasures. The ECM-ECCM battle is open-ended and will continue indefinitely into the future. The sensors on the newest missiles that are entering into the operational inventories of potential adversaries appear to have extremely robust ECCM capabilities against current ECM techniques. Advances in techniques related to automatic target recognition (ATR) tend to make seekers robust against distraction decoys. On the other hand, the fidelity with which modern decoys or repeaters can replicate the signature of the target of a cruise missile is impressive. Clearly, the Navy must continue an aggressive ECM program so that as new advances in seeker technology are fielded, new countermeasures will be available to negate them.

Aside from the threat that improved cruise missile seekers pose to Navy ECM techniques, there are many trends in cruise missile design that the committee found to be a source of concern, including the following:

- Greater missile speeds, which limits the engagement time;
- Lower RCSs, which limits the range within which a missile may be detected once it has crossed the horizon of defensive radars;
- High maneuverability, which limits the ability of a defensive system to track and engage the missiles;
- Trajectories that make maximum use of terrain obscuration and clutter masking in littoral situations; and
- Sea-skimming flight paths, which keep incoming missiles below the horizon of defensive radars for as long as possible.

Worldwide, cruise missile designs abound. Many of these designs already stress the capabilities of U.S. defensive systems. Table 2.1 lists some of the worst-case parameters of currently operational missiles and the committee's projections for the parameters that may be encountered in the next 15 to 20 years, based on its assessment of trends in technology.

The first four attributes listed in Table 2.1 are intended to limit the options for engagement by defensive missiles. The fifth and sixth attributes attempt to defeat defensive ECM techniques. The committee's estimates for 2020 are extrapolated from current trends in missile technology. The sixth could leapfrog future ECM efforts. Although some members of the committee doubt that accelerations of 20+ *g* will be feasible, all of them concur that future cruise missiles will possess greater agility than currently deployed threat missiles. As missiles become more agile, the data rates of defensive systems must be increased and the track association algorithms of defensive sensors will require major modifications. Even if the RCS values of threat missiles do not decrease, their greater agility and speed will challenge the tracking algorithms and data rates of existing defensive systems.

TABLE 2.1 Attributes of Current and Projected (2020) Cruise Missile Threats

| Attribute | Current | Estimated for 2020 |
|---|---|---|
| Agility | 6 to 8 *g* maneuvers | 10 to 20+ *g* maneuvers |
| Nose-on radar cross section | –10 to –30 dBsm | –20 to –40 dBsm |
| Altitude (sea skimmers) | 2 to 5 meters | 2 to 5 meters |
| Terminal speed | Sub- to supersonic | Up to hypersonic |
| Electronic countermeasure robustness | Moderate | Improved |
| Guidance | Global positioning system (GPS) and radar/infrared | GPS and target recognition |

Under optimum conditions, these systems employ a defensive shoot-look-shoot (SLS) doctrine to conserve interceptors. If the defensive time line is compressed by some combination of changes in missile RCS altitude, ECCM, speed, and maneuverability, the capability of the defensive system will be reduced successively from shoot-look-shoot, to shoot-shoot, to shoot and, in the worst case, to no shot possible. The effects of RCS and Mach number are illustrated conceptually in Figure 2.1.

The figure shows that for a sea-skimming incoming cruise missile attacking a defended ship equipped with only a surface-based sensor (or for such a missile at any given altitude), there will be a large area in the RCS-missile velocity plane where it is not possible to launch a defensive round. If the altitude of the attacking cruise missile is low enough and if its velocity is high enough, there may be no way to shoot the missile, even if it has a large RCS. In other areas of the RCS-missile velocity plane, defensive systems have an opportunity to launch either one or two defensive missiles or may even be unable to launch a single defensive missile.

Most current cruise missile threats do not lie in the no-shot region. However, unless elevated sensors are used, or unless significant improvements in defensive capabilities are achieved, missiles with the attributes in the third column of Table 2.1 will generally fall into the no-shot region of Figure 2.1. In such a situation, the defense will have to depend entirely on the Navy's ECM capabilities to defeat the terminal guidance system.

If elevated radars are used in lieu of surface-based radars, the radar horizon will increase significantly, somewhat negating the advantages of high speed. Of course, elevating the radar will not offset a reduction in the RCS value of the threat missile. The RCS value of a missile varies with both frequency and missile orientation. Thus, low-RCS missiles can only be defeated by using radars that operate at lower radar frequencies and/or by using some form of

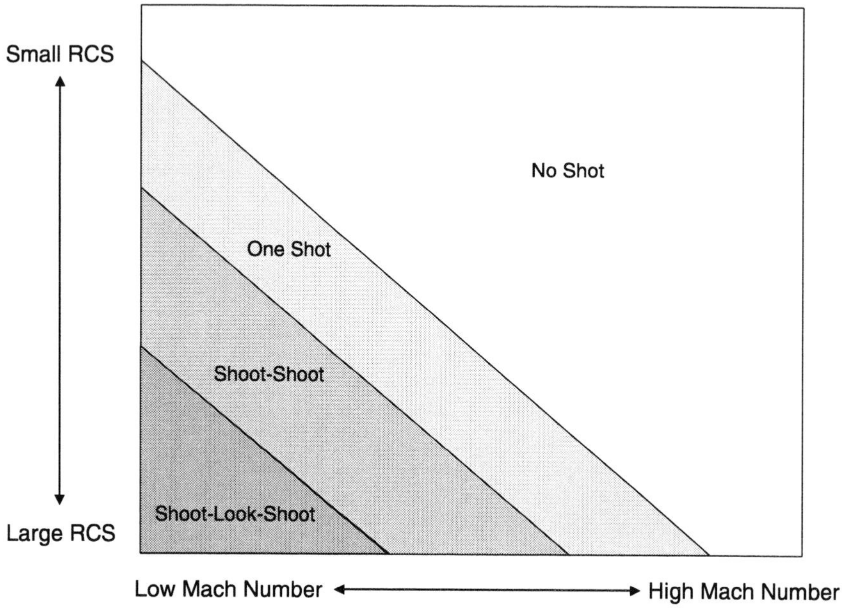


FIGURE 2.1 Conceptual representation of the effects of radar cross section (RCS) and Mach number on the defensive capabilities of a surface-based sensor and associated missile system against a cruise missile attacking at a sea-skimming altitude.

multistatic operation that allows the missile to be viewed from orientations where its RCS value has significant peaks.

## 2.3 THEATER BALLISTIC MISSILE THREATS

### 2.3.1 Characteristics of Theater Ballistic Missiles

On September 26, 1997, agreements were signed between representatives of the United States and the Russian Federation that established the permissible characteristics of TMD systems.

The 1st Agreed Statement permits either side to deploy lower-velocity TMD systems (those with interceptor speeds below 3 km/s) provided that they have not been tested against a ballistic missile target having a range greater than 3500 km or a speed greater than 5 km/s. (U.S. compliance review has independently determined that PAC-3, THAAD, and NAD systems are compliant with that agreement.)

The 2nd Agreed Statement permits either side to test interceptors that are faster than 3 km/s against ballistic missile targets with velocities less than 5 km/s and ranges less than 3500 km. (The Navy's upper-tier TMD has been certified as being compliant with the ABM Treaty.) Strictly speaking, the 2nd Agreed

Statement only says that a requirement for being legal under the ABM Treaty is that such higher speed TMD systems must not be tested against targets faster than 5 km/s. However, this does not automatically mean they are compliant. With this delineation, one can define theater ballistic missiles (TBMs) as one- to three-stage liquid- or solid-propelled rocket vehicles that have launch-to-impact ranges of 100 to 3500 km. Although these limitations have not been submitted to, or ratified by, the Senate, the Clinton administration adopted them as policy.

From first principles, TBM velocities at final-stage burnout must be from 1 to 5 km/s, while their post-boost times of flight vary from about 2 to 20 min (see Figure 2.2). After the powered phase of flight, a TBM flies in a vertical

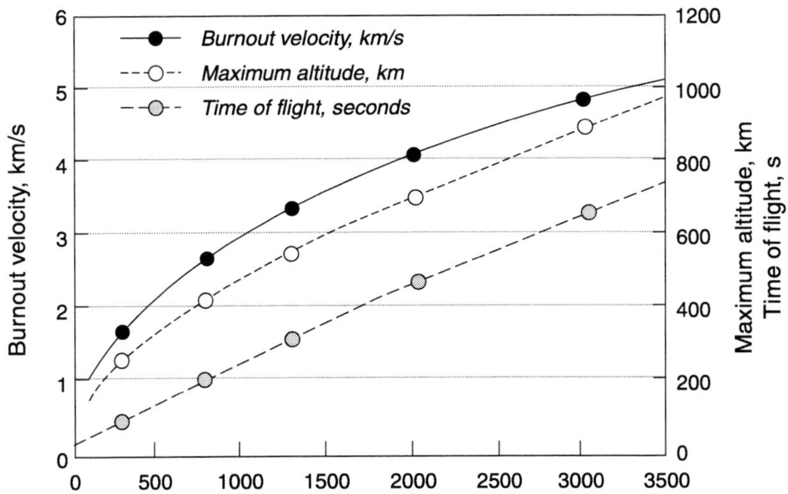


FIGURE 2.2 Minimum burnout velocity, maximum altitude, and time of flight as functions of range between burnout and reentry. Burnout altitude = reentry altitude = 10 km with Earth's rotation and atmospheric effects neglected. NOTE: This figure is approximate and actual values will vary according to more detailed scenarios. SOURCE: Calculations based on (1) Bate, Roger R., Donald D. Mueller, and Jerry E. White, 1971, *Fundamentals of Astrodynamics*, Dover Publications, Mineola, N.Y.; Thomson, W.T., 1961, *Introduction to Space Dynamics,* Dover Publications, Inc., Mineola, N.Y., December; (2) Thomson, William T., 1961, *Introduction to Space Dynamics*, Dover Publications, Mineola, N.Y., paperback edition, May 1986; (3) Sellers, Jerry J., and Wiley J. Larson, 1961, *Understanding Space: An Introduction to Astronautics, Space Technology Series*, McGraw-Hill, New York, N.Y.; and (4) Sellers, Jerry J., William J. Store, Robert B. Giffen, and Wiley J. Larson, 2000, *Understanding Space: An Introduction to Astronautics, 2nd edition,* McGraw-Hill College Division, McGraw-Hill, New York, N.Y., June.

plane between the launch site and the target. The time of flight and distance traveled during boost, a function of acceleration as well as staging, typically are small percentages of their respective totals.

The corresponding maximum altitudes vary from about 20 to about 800 km depending on impact range and trajectory type (maximum-range, depressed, or lofted). A missile can be targeted at less than its maximum range by depressing or lofting the trajectory or by reducing the burnout velocity. Depressed trajectories to a given impact range have shorter flight times and maximum altitude. Their shallow reentry angles tend to degrade impact point accuracy unless they are terminally guided. Average heating and aerodynamic loads may be increased as a consequence of more time spent in endo-atmospheric flight.

There is no clear demarcation between atmospheric and nonatmospheric flight because air density decreases exponentially with altitude. Nevertheless, above a certain altitude, typically 70 to 80 km, aerodynamic forces on the TBM are negligible because the dynamic pressure (one-half of the air density times velocity squared ($P_{dynamic} = ½ P_{air} \times V^2$) is low. Flight above this altitude is called exo-atmospheric flight, while flight at lower altitudes is called endo-atmospheric (or simply atmospheric) flight. Exo-atmospheric flight is unlikely for an impact range of less than 150 km unless the missile follows a lofted trajectory. For a greater impact range, the exo-atmospheric flight time varies from 0 to less than 20 min, while the endo-atmospheric flight time following reentry is 10 to 40 s. The boost phase may extend into the exo-atmospheric regime for an impact range of more than 300 km.

Control forces for midcourse guidance correction can be provided by thrust during exo-atmospheric flight, while terminal guidance corrections can be effected by aerodynamic lift and drag control during endo-atmospheric flight. Sources for measuring guidance error—a necessary input to the guidance logic—include inertial measurement units; GPS or GLONASS; and terminal homing sensors, such as laser designation, optical or radar imaging and distance measurement, and radio stations or beacons.

Each TBM has one or more warheads whose size and mass are dependent on the payload capability of the launch vehicle. Unitary warheads are likely to contain a high explosive—including the nuclear alternative—to damage or destroy the target, although a precisely guided inert penetrator could be considered for attacking a deeply buried asset. Multiple warheads, including independently targeted RVs or unguided submunitions, could carry explosives for damaging one or more targets or chemical or biological agents to attack personnel over a wide target area; for this reason, their effectiveness is less dependent on impact-point accuracy. Multiple warheads could be released shortly after boost, or they could be dispensed from a guided or unguided bus (carrier) vehicle following reentry.

Even if TBM warheads are successfully engaged after boost, the remnants proceed on a ballistic path toward the vicinity of the target. If substantial destruction has been accomplished, the components or fragments are likely to have a reduced ballistic coefficient, causing them to fall short of the target. For exo-atmospheric engagement, the parts could be further damaged by reentry heating, especially for longer range (and therefore higher velocity) missiles. Following endo-atmospheric intercept, the pieces—possibly including fully functioning submunitions—may still rain down on the target.

Since atmospheric heating increases the temperature of the TBM warhead during boost and reentry, the warhead is detectable in both exo- and endo-atmospheric flight not only as a radar target but also as an infrared source.

Countermeasures are intended to mask the position of the TBM warhead(s) from defensive sensors or to evade a defensive weapon. Spent boosters, debris fields (e.g., from a detonated booster that is not too close or too far), and deployed decoys may be difficult to discriminate from warheads, while the warhead itself could be cooled to make it less visible to infrared sensors. Given a field of incoming targets whose radar or infrared signatures are similar, defensive sensors must pick out the right targets for attack. The latter approach adds to the actual dispersion of the trajectory, and it increases the physical difficulty of killing the warhead, though TBM impact-point dispersion still could be contained by terminal guidance.

The time to deployment for specific theater ballistic missiles is a critical issue. Near-term, mid-term, and long-term TBM threats must be considered, and there is a big distinction between an actual threat and a possible threat. As the window of concern lengthens, that is, as the planning horizon for TBM defense stretches out, the current ("actual") becomes less interesting and the "possible" becomes more real, especially for technologies whose development can be kept from surveillance by intelligence. Therefore, programs for future defense against TBMs must take into account not only the characteristics of known threats but also the technologies that can be employed in response to an adversary's perception of our nation's defensive capabilities.

Scuds and their derivatives, which thus far have accounted for the bulk of widely proliferated missiles, are generally inaccurate and do not separate spin or reorient their payloads.

### 2.3.2 Current Theater Ballistic Missile Threats

For the present, it is sufficient to consider the TBMs discussed in NAIC-1031-0985-98 (4) as representative of current threats.[1] While many of these missiles embody old and relatively rudimentary technology, the more sophisticated missiles that are replacing them in the inventories of supplier nations today

[1]National Air Intelligence Center. 1998. "Ballistic and Cruise Missile Threat." Wright-Patterson Air Force Base, Ohio. Available online at <http://sun00781.dn.net/irp/threat/missile/naic/index.html>.

can be expected to proliferate in the same way as the older systems did—by being exported.

Figure 2.3, compiled from various unclassified sources, illustrates the major ballistic missile threats that the naval theater missile defense systems must contend with.

The simplest TBMs are unguided beyond the boost phase, after which they follow ballistic trajectories toward their intended targets without trajectory corrections. They tend to be subject to large dispersions in impact point as a consequence of uncertain burnout conditions, physical modeling errors, tumbling and coning motions, and variations in wind and air density during endo-atmospheric flight. More sophisticated TBMs employ the following means to reduce reentry dispersion:

- Separation of the TBM warhead from the launch vehicle in order to effectively increase the ballistic coefficient (mass divided by drag coefficient and reference area),
- Reorientation of the RV to minimize the angle of attack (angle between the vehicle's axis of symmetry and the velocity vector) at reentry, and
- Provision of spin to the RV to provide gyroscopic stability and to cancel the trimmed lift force during and following reentry.

Control forces for midcourse guidance correction can be provided by thrust during exo-atmospheric flight, while terminal guidance corrections can be effected by aerodynamic lift and drag control during endo-atmospheric flight. Sources for measuring guidance error—a necessary input to the guidance logic—include inertial measurement units, GPS or GLONASS, and terminal homing sensors, such as laser designation, optical or radar imaging and distance measurement, and radio stations or beacons.

### 2.3.3 Postulated Future Theater Ballistic Missile Threats

Threat missiles will become more sophisticated in the coming years.[2] Improved accuracy for warhead delivery and some form of countermeasures are almost certain to be incorporated into the TBMs that U.S. forces will face. The

[2](1) National Air Intelligence Center. 1998. "Ballistic and Cruise Missile Threat," Wright-Patterson Air Force Base, Ohio. Available online at <http://sun00781.dn.net/irp/threat/missile/naic/index.html>; (2) Committee on Foreign Relations. 1999. "Foreign Missile Developments and the Ballistic Missile Threat to the United States Through 2015," Hearings Before the Committee on Foreign Relations, U.S. Senate 106th Congress, First Session, U.S. Government Printing Office, Washington, D.C., April 15 and 20, May 4, 5, 13, 25, and 26, and September 16. Available online at <http://www.fas.org/spp/starwars/congress/1999_h/s106-339-8.htm>; (3) Commission to Assess the Ballistic Missile Threat to the United States. 1998. "Executive Summary of the Report of the

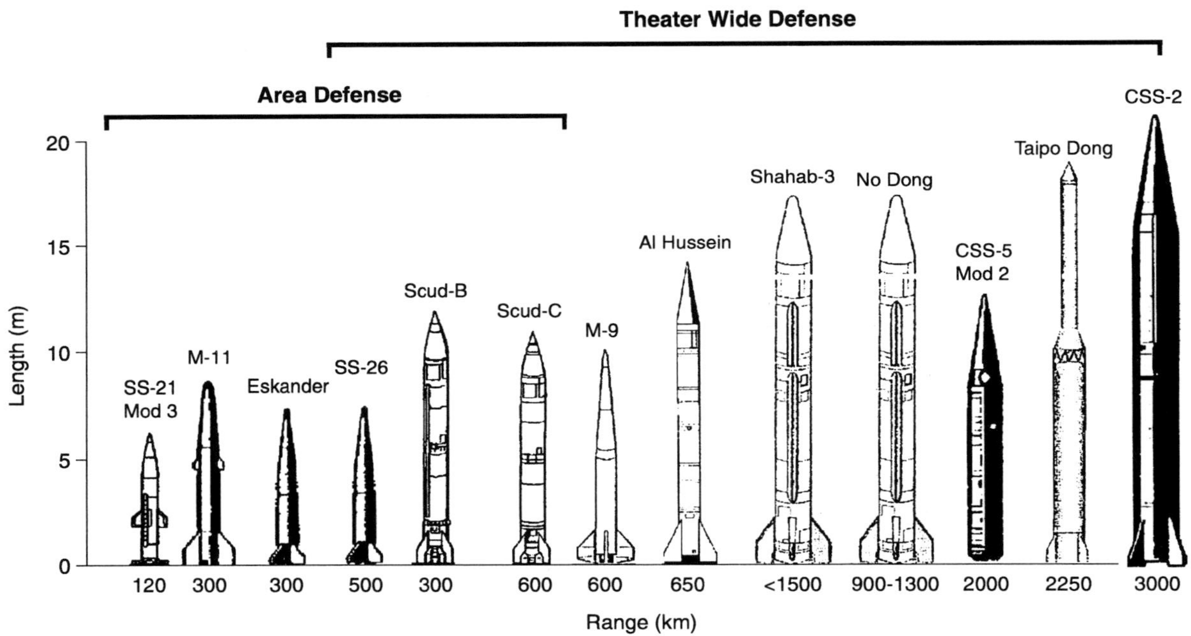


FIGURE 2.3 Tactical ballistic missile threats and their characteristics. SOURCE: Information extracted from (1) Cosby, Anthony W., "Protection from the Missile Threat," (Unclassified) May 24, 2000, briefing to the committee, Program Executive Office, Air and Missile Defense, Arlington, Va.; (2) Patterson, CDR Sheila A., USN, "Navy Theater Wide TBMD Program (U)," (Classified) June 28, 2000, briefing to the committee, Program Executive Office, Theater Surface Combatants (PMS 452), Arlington, Va.; and (3) Rempt, RADM Rodney P. USN, "Naval Theater Missile Defense for the 21st Century," (Unclassified) May 24, 2000, briefing to the committee, Office of the Deputy Assistant Secretary of the Navy for Theater Combat Systems, Washington, D.C.

threat indicators are already present. Newer TBMs incorporating accurate warhead delivery are in (or will soon enter) Russian and Chinese inventories.

No one can state with certainty what specific countermeasures will be incorporated into threat TBMs within the next 5 to 20 years. However, the committee suggests that future threat TBMs, in response to the presence of defense systems, might incorporate some combination of the following capabilities so as to stress the capabilities of present and planned TBMD systems:

- RVs with reduced RCS,
- Flares and or IR chaff,
- Radio-frequency chaff,
- Escort jammers,
- Decoys and/or tethered objects,
- Shrouds to mask IR signatures,
- Coated boosters that are robust against laser attack, and
- Deceptive maneuvering.

Some of these techniques (chaff, jammers, low-RCS RVs, shrouds, and so on) would stress the ability of our military's primary and secondary target acquisition sensors to detect and track the RV of interest. Others pose a sensor discrimination problem—for example, How does a TBMD system differentiate between an RV and a decoy? The committee believes that the situation in TBMD is much like the competition between ECM and ECCM techniques in the ASCMD arena. Although the foregoing potential countermeasures to a TBMD system are a significant concern, none of them is inherently immune to negation.

The committee takes note of the vigorous debate that rages about exo-atmospheric discrimination and the ease of creating effective countermeasures. The

Commission to Assess the Ballistic Missile Threat to the United States," U.S. Government Printing Office, Washington, D.C., July 15. Available online at <http://www.fas.org/irp/threat/bm-threat.htm>; (4) Institute for Foreign Policy Analysis. 1997. "Exploring U.S. Missile Defense Requirements in 2010: What Are the Policy and Technology Challenges?" Washington, D.C., and Cambridge, Mass. Available online at <http://www.fas.org/spp/starwars/advocate/ifpa/>; (5) APS Forum on Physics and Society. 1994. *Symposium on Theater Ballistic Missiles: What Is the Threat? What Can Be Done?* American Physical Society, held in Washington, D.C., on April 18, published as Vol. 23, No. 4, October. Available online at <http://www.positron.aps.org/units/fps/aoct94.html>; (6) Director, Operational Test and Evaluation. 2000. "Navy Area Theater Ballistic Missile Defense (NATBMD)," *DOT&E FY99 Annual Report to Congress,* Department of Defense, Washington, D.C., February. Available online at <http://www.dote.osd.mil/reports/FY99/other/99natbmd.html>; and (7) Sessler, Andrew M., John M. Cornwall, Bob Dietz, Steve Fetter, Sherman Frankel, Richard L. Garwin, Kurt Gottfried, Lisbeth Gronlund, George N. Lewis, Theodore A. Postal, and David C. Wright. 2000. *Countermeasures: A Technical Evaluation of the Operational Effectiveness of the Planned U.S. National Missile Defense System,* MIT Security Studies Program, Union of Concerned Scientists, Cambridge, Mass., April. Available online at <http://www.ucsusa.org/security/CM_exec.html>.

contest between countermeasures and counter-countermeasures often hinges on particular idiosyncrasies or "tags" that can be exploited by one side or the other. Knowing how potential adversaries are exploiting their tags allows denying the tags. It is observed that those actually working on the defeat of countermeasures must of necessity keep silent, while those who raise issues of the difficulties of defeating countermeasures on the basis of so-called "physical first principles" arguments are not hampered by security issues. With insight into some of the ongoing restricted or classified work in this area, the committee would caution against the oversimplistic arguments often heard in the public rhetoric.

While easily postulated, many countermeasures would be difficult to achieve. For example, many have proven to be difficult for U.S. engineers to incorporate into U.S. missiles. The committee wishes to emphasize that once deployed, a TBMD system must be upgraded periodically in response to observed threat indicators. Therefore, programs for future defense against TBM missiles must take into account not only the characteristics of known threats, but also the technologies that an adversary can employ in response to its perception of our defensive capabilities. Techniques to negate the countermeasure threats listed above may take several years to develop and implement. The committee believes that a robust and sustained R&D program to develop specific naval TMD upgrades to negate those techniques should be in place, and it addresses this point in Chapter 4.

For TBMs without an attitude control motor (ACM) or bus, the attacker has limited flexibility in the type of penetration aids carried and their deployment. Objects need to be deployed either from the booster or from the RV, and the attacker must be sure that the penetration aides cover the RV but do not give away its location. There are further constraints on the orientation of the RV, which may impose more severe requirements on the penetration aids, increasing their weight or decreasing their numbers.

For TBMs with an ACM, the attacker has considerably more freedom in the design and deployment of penetration aids. If the RV's orientation can be controlled, the attacker can take advantage of RCS reduction, which significantly decreases the radar detection range and makes it much easier to use maskers such as chaff and jamming.

Warhead maneuvers, intentional or not, increase the difficulty of intercept. Thus, although a tumbling TBM may be relatively easy to detect, hitting its warhead still may be a challenge. Relatively small thrust impulses applied during exo-atmospheric flight can induce spiraling motions in oblong or dumbbell-shaped bodies, while aerodynamic forces and moments produce spiraling or jinking of a streamlined body during reentry. The amplitude and frequency of such maneuvers can have a first-order effect on an interceptor's ability to sense and engage the warhead. Both deception and maneuvering can increase the uncertainty in estimates of the TBM trajectory. It must be recognized that both approaches add to the actual dispersion of the trajectory and that neither is compatible with the accuracy needed for conventional TBM unitary warhead target damage objectives. With penalties in payload complexity and displacement of warhead volume and weight, terminal guidance and maneuver capability could be added to maintain acceptable impact-point dispersion. The emergence of that capability could be a threat indicator for these induced-motion types of countermeasure.

# 3

# Assessment of the Current State of Technologies Involved in Naval Theater Missile Defense and an Evaluation of Current and Projected Department of the Navy Programs Designed to Meet the Evolving Threat

This chapter assesses the Department of the Navy's current and projected capabilities in theater missile defense and the status of the technologies involved. It begins with a summary evaluation of the Navy's and Marine Corps' overall current and projected capabilities in the three distinct missions: ASCMD, OCMD, and TBMD. Then the discussion turns to subsystems in order to focus on technology, treating, in turn, sensors, weapons, and BMC3 systems.

Central to the effective utilization of these technologies are concepts of operation for executing the missile defense missions. The committee sought, during several of the Navy and Marine Corps briefings, to understand the concept of operations that would be used in the conduct of expeditionary operations. In particular, the committee wished to learn how the theater missile defense operations might be coordinated with the other operations that would be taking place at the same time and in the same area. Various presentations indicated that aircraft would be operating to deliver and provide logistic support to Marine Corps units ashore and that fire missions would be executed by ships launching ERGMs and other land-attack weapons, as called for by the Marine Corps. The committee believes it is necessary to construct a concept of operations that uses whichever measures are necessary to ensure that the theater missile defense can be coordinated with the offensive operations in such a manner that both succeed without conflict or danger to friendly forces. The briefers were unanimous in the opinion that no such concept has yet been defined.

Joint doctrine has been promulgated to guide the conduct of air and missile defense in a theater;[1] however, the doctrine appears to presume that the theater has already been developed and that joint forces are in place. Other than to note that the same functions must be performed in undeveloped theaters, the joint doctrine is not helpful as a guide for expeditionary warfare. Although work is ongoing to evolve this doctrine at the Joint Theater Air and Missile Defense Organization (JTAMDO), the committee is not aware of any efforts to address the expeditionary warfare setting.[2]

Pending the development of doctrine to guide initial operations in an undeveloped theater, it falls to the Navy and Marine Corps to define the appropriate CONOPS. A CONOPS for expeditionary warfare in the littorals must address conflicting requirements for employment of operational assets and for control of offensive and defensive operations.

Concepts for conduct of the offense are amenable to preplanning to avoid conflict yet must remain flexible enough to support operations ashore by Marine Corps units that may become subject to variation because of real-time events. At the same time, and in the same area, defensive measures must be taken to defeat ballistic missile, cruise missile, and aircraft threats to forces in the area, both afloat and ashore.

The conduct of effective theater missile defense without disruption of and conflict with offensive measures is a very difficult task but a necessary one. However, several briefers told the committee that no concepts for coordinating offensive and defensive operations have been worked out. Developing such concepts is critical to the conduct of expeditionary warfare and deserves considerable effort. Such concepts are also necessary to a proper evaluation of the adequacy of theater missile defense programs.

## 3.1 OVERVIEW OF THEATER MISSILE DEFENSE CAPABILITY

As discussed briefly in Chapter 1, the Department of the Navy's mission, which is to operate in the littorals and influence events ashore, has a strong impact on TMD requirements, and—as discussed in Chapter 2—the air threat

---

[1]Fulford, LtGen C.W., Jr., USMC, Director, Joint Staff. 1999. "Joint Doctrine for Countering Air and Missile Threats," Joint Publication 3-01, The Pentagon, Washington, D.C., October 19. Available online at <http://www.dtic.mil/doctrine/jel/new_pubs/jp3_01.pdf>; Ross, Lt Gen Walter K., USAF, Director, Joint Staff. 1996. "Doctrine for Joint Theater Missile Defense," Joint Publication 3-01.5, The Pentagon, Washington, D.C., February 22. Available online at <http://www.dtic.mil/doctrine/jel/new_pubs/jp3_01_5.pdf>.

[2]Joint Theater Air and Missile Defense Organization and Ballistic Missile Defense Organization. To be published. "Annex G: GTAMD 2010 Operational Concept (Draft Version 5 (Unclassified—For Official Use Only))," from *1999 Theater Air and Missile Defense (TAMD) Master Plan (U),* The Pentagon, Washington, D.C., December 2 (Classified).

continues to become more difficult. This section discusses the effect of these factors on overall naval capabilities in ASCMD, OCMD, and TBMD.

Because antiship missiles in the hands of potential adversaries are so numerous, so sophisticated, and so widespread, and because every naval combatant becomes a target whenever it enters the theater and must defend itself well so as to be an asset rather than a liability, ASCMD must be the Navy's highest priority in TMD. While the Navy's current capabilities are inadequate against antiship cruise missiles and its funding plans insufficient to protect some classes of ships against them, the Service has in hand the fundamental framework for effective defense against foreseeable ASCM threats.

However, the Department of the Navy has not come to grips with the rapidly approaching necessity for overland cruise missile defense. In the future, adversaries will employ land-attack cruise missiles to deny U.S. forces needed access to ports and airfields in theaters of war. In the fundamental framework for an OCMD system, important elements are missing.

Because tactical ballistic missiles are widespread weapons of terror and potential mass destruction and are poised today to deny U.S. access into theater, the nation needs, as soon as possible, a capability that will provide TBMD for ports and airfields until assets of other Services are in place. The Navy's burgeoning TBMD capability divides into two parts: area- and theater-wide systems. There are clear differences in how well the two systems are progressing toward an effective operational capability. The NAD system promises robust local defense against the short- to medium-range TBMs prevalent today and appears to be progressing smoothly. The NTW system, on the other hand, is a demonstration, not an acquisition program. The activity lives year to year on funding provided through congressional plus-ups. Its initial capability, if indeed it becomes a funded acquisition program and the Program Executive Office for Theater Surface Combatants' current plans for it continue, will be limited by SPY-1 radar performance, which in some geographic scenarios is inadequate to provide the wide defensive coverage needed to deal with the threat of ever-longer-range TBMs. Furthermore, since adequate TBMD requires defense in depth, a Navy theaterwide capability will one day be required.

The next subsections delve further into the three mission areas (ASCMD, OCMD, and TBMD).

### 3.1.1 Antiship Cruise Missile Defense

For the past half century or more, naval battle groups have been defended in several layers. The outermost layer has been air-to-air combat, a capability to "shoot the archer." In recent decades, Aegis ships have provided the second layer, an umbrella of area defense over the battle group. Once the most important layer in battle group defense in depth, area defense now yields primacy to self-defense, largely owing to the severity of low-altitude threats. Nevertheless, area defense can help in important situations such as the close-escort protection of aircraft carriers.

Beginning with the AAW capstone requirements document (CRD) in 1996, the Navy has characterized the AAW performance of various ship classes in terms of a "probability of raid defeat," whereby a raid is considered to be defeated if no threat missile penetrates the defense to hit the ship. The Navy defines the "probability of raid defeat" as a weighted sum of results against a specific raid (e.g., *x* low-altitude, low-observable, subsonic cruise missiles in *y* seconds). The CRD varies the *x* and *y* and the required "probability of raid defeat" by ship class. The committee takes no exception to the numbers in the CRD. The weightings are done across different classes of threat. A present-day low-altitude, low-observable, subsonic cruise missile is an example of class. The CRD does not specify the weightings. The Navy practice has been to give a heavy weight to the moderate cruise missile threats predominant today and much less weight to the more difficult threats, which are expected to emerge in the future or—if they already exist—are less numerous. This weighting tends to have a stronger effect on ship classes with less stringent requirements.

The Navy justifies the CRD requirements and the weighting practice as a way to allocate scarce funds, because it cannot afford to defend all ships equally. This is no doubt so. However, the committee fears that such a practice tends to obscure real vulnerabilities. The adversary will decide which ship to attack and with what missiles. The adversary may "win" (if, for example U.S. popular opinion turns against further action) by attacking and sinking a less-well-defended ship with the best cruise missile it can buy. Some ship classes will not have to operate for long periods of time in the littorals and be exposed to the full threat, but others will. The committee believes that any ship so exposed should have the benefit of the best defense the Navy can provide.

In the past, the air cover provided by Aegis ships was effective over a large area. Self-defense systems on some ship classes lagged in capability, but robust area defense gave Navy battle groups a good overall AAW capability. Today, with typical ship formations, the ability of one ship to defend another against some of the most dangerous antiship cruise missiles is almost nil, because the threats fly too low and too fast.

As the information presented to the committee by a number of Navy offices clearly shows, the Navy's overall current operational capabilities in antiship cruise missile defense are marginal and declining. In recognition of this, the Navy has been investing heavily in a number of new detection, control, and engagement systems and also in systems integration. When these new capabilities are fielded, antiship cruise missile defense will be markedly improved for the ships that receive them.

The committee is confident that the Navy has in hand the framework for antiship cruise missile defense. It consists of a combination of volume-search and horizon-search radars, well-automated fire control and doctrine that permits fast response through automated decision-making in high-threat situations, defensive missiles that match the sophistication of the threat, and a netted capability that enables distributed ships to fight as a coherent whole. These elements, fielded on the right ships in adequate numbers and in timely fashion, should enable the Navy to counter foreseeable antiship cruise missile threats. Ships that do not receive these upgraded capabilities will remain vulnerable. The current program of record does not fully deploy the new capabilities. For example, it appears that ships other than Aegis cruisers and destroyers will lack an adequate engagement capability. A launcher for the ESSM will not be available on these vessels, and they must depend on the rolling airframe missile (RAM). Nor does the current program of record field capabilities to cover all the potential electronic countermeasure threats. The committee believes that the Department of the Navy should prioritize funding so that every combatant that conducts sustained operations in contested littoral waters is adequately defended.

In summary, providing adequate defense against antiship cruise missiles will require reprioritization of funding, but the fundamental framework for ASCMD is there.

### 3.1.2 Overland Cruise Missile Defense

As mentioned earlier, the committee believes that the Department of the Navy has not come to grips with the rapidly approaching necessity for an overland cruise missile defense. Important elements are missing from the fundamental framework for such a system.

In the past, the Marines carried improved Hawk batteries into the theater for overland air defense, primarily against aircraft threats. Then, in the interest of mobility, they retired this improved system, which was bulky. The Marine Corps is now developing a point defense capability, but for the foreseeable future, U.S. forces entering the theater will have no wider defense coverage until the Army's Patriot batteries can be put in place.

The Marine Corps operational strategies, OMFTS and STOM, will require the Navy to provide layers of air defense overland. Carrier-based manned aircraft can be counted on to keep enemy manned aircraft at bay, but in the future the enemy may use land attack cruise missiles to attack fixed objectives such as ports and airfields. As discussed in Chapter 2, land attack cruise missiles are not common today, but the nation's current weaknesses in countering them may hasten their development and deployment.

If Navy platforms are to provide an overland cruise missile defense, there must be a capability to detect, track, and intercept cruise missiles that are beyond the line-of-sight horizon of ships at sea. One possible operational concept for OCMD includes an airborne platform for detection, weapon launch from a surface ship, in-flight control by the ship based on the airborne platform's track ("engage on remote"), and active terminal guidance by the weapon. Alternatively, the airborne platform could control the weapon in flight ("forward pass") or it could illuminate the target for semiactive homing by the weapon. Whichever concept is considered, major elements are missing. The Navy has no airborne platform capable of detecting a low-observable cruise missile overland. It has no ship-launched, actively guided air defense weapon. It has no airborne illumination capability.

The Department of the Navy is unprepared for a defense against land-attack cruise missiles and is not funding development to rectify the situation.

### 3.1.3 Theater Ballistic Missile Defense

The NAD system will implement a TBMD capability on all Aegis cruisers and destroyers. The NAD system requires changes to the SPY-1 radar, to the Aegis weapon control system, and to the standard missile (SM). The system is being designed to defend a limited region around the ship against short- to medium-range TBMs. The reach of the system will enable ships to operate a few tens of miles offshore and defend assets a few tens of miles inland. While the SPY-1 radar will be taxed, improvements under way in the NAD program should enable it to detect TBMs at ranges matched to its interceptor's kinematic range. Engagements with the NAD system will occur well within Earth's atmosphere, and atmospheric drag will strip away much of the confusing debris around the TBM warhead, simplifying the target discrimination problem. The NAD interceptor, denoted SM-2 Block IVA, employs the propulsion stack of a currently operational SM-2 variant and adds an IR guidance system, among other things. It operates deep in the reentry region and uses aerodynamic maneuvering. This region is where threat RVs may also maneuver either inadvertently or deliberately. Short-range TBMs have low velocities and cannot maneuver very strongly. The high-*g* capability of the SM-2 together with its warhead should give it reasonable single-shot or salvo capability against these targets.

The NAD system, as described to the committee, appears to be well structured and, except for the inadequate funding for the spiral development evolution, appears to have a well-defined development path that is supported by good analytic underpinnings. The system strengths and limitations are well understood and are being treated appropriately. The NAD system objectives for tactical ballistic missile defense are realistically limited and clearly stated. The performance of the system against its design threats was presented clearly and not overstated. The area TBMD challenge is a formidable one, and the Navy and DOD should probably expect some setbacks in the course of development, but the conceptual design and the program to develop it appear sound.

Like the NAD program, the NTW effort intends to build on the Aegis legacy. However, the longer-range TBMs the NTW system is intended to counter and the much broader areas it is intended to protect place a far greater burden on the system. The NTW system will employ a highly modified standard missile

(SM-3), with the original warhead/seeker stage replaced by a new third-stage motor and a HTK vehicle. With a much lighter final payload, the interceptor burnout velocity is much greater, permitting much longer fly-outs and thus much larger defended regions. To get the large coverage, the SM-3 is designed to intercept exo-atmospherically, which necessitates launching the interceptor much earlier in the threat missile trajectory. The weakest link in the proposed phase I NTW defense is the detection capability the Navy will obtain by evolutionary improvement of Aegis's SPY-1 radar. In geographic situations where the NTW ship can be placed near the TBM launch point, the protected region can be very large. However, in situations where the NTW ship is near the TBM aim point, the protected region can be very small, limited as it is by SPY-1's detection capability.

The committee believes that, certainly in NTW phase I and probably beyond, the Navy must devise concepts of operation that take advantage of detection assets not organic to the battle force. Forward-placed or space-based assets that detect TBMs early in flight would, through CEC or a similar link, enable midcourse control of NTW interceptors in order to greatly increase the size of the defended footprint in unfavorable geographies.

The Navy is considering a new generation of shipboard radars for TMD. Achieving adequate detection and discrimination for NTW ships will be a driving requirement. One concept combines an S-band volume-search radar (much more powerful than SPY-1) with an X-band radar for horizon-search against cruise missiles and for long-range TBM discrimination.

Because the severity of the near-term threat calls for fielding an NTW capability quickly, because many engineering challenges must be overcome to field even a limited NTW capability, and because the Navy will surely benefit considerably from experience gained in beginning to use the system as soon as possible, the Navy is considering fielding the so-called Block I NTW system. It is clear, however, that during the years it will take to field the system, the TBM threat will become even more severe, especially in the use of penetration aids, partly in response to the advent of the NTW system itself. The Navy's informal plans call for a Block II capability against a more severe threat, but the R&D to solve the challenges Block II will face is dragging. The committee also believes that in some geographic scenarios, the NTW system may ultimately need to depend on detection capabilities not organic to the ship in order to achieve wide defensive coverage.

## 3.2 SUBSYSTEM TECHNOLOGY ASSESSMENT

The next subsections assess the state of technologies in the subsystems employed in the Department of the Navy's TMD. The technologies involved in sensor and weapon subsystems are the primary focus, but weapon control and electronic warfare are also discussed.

### 3.2.1 Sensors

Because requirements and technologies are so diverse in different parts of the theater missile defense realm, the discussion of sensors is divided into assessment of technologies for (1) shipboard TBMD sensors, (2) National sensors for TBMD, (3) sensors for overland cruise missile defense, (4) sensors for anti-ship cruise missile defense, (5) sensors for air-to-air combat, and (6) sensors for electronic warfare.

#### 3.2.1.1 Shipboard TBMD Sensors

The role of the sensors in a TBMD system is to detect, locate, track, and identify the RV and to provide information that will permit an interceptor to hit it. This section focuses on the pre-weapon-commitment sensor, which is generally a surface-based radar that may be augmented by space or airborne IR sensors. The sensors used by interceptor seekers are discussed in Section 3.2.2.1, "TBMD Weapons."

Ballistic missiles generally arrive from high altitudes at high angles of elevation. Thus, detection range rather than terrain masking or clutter is generally the limiting factor in the TBMD performance of a surface radar. Once a nonmaneuvering RV has been detected, its probable impact area is easily determined. Generally, the problem of distinguishing friend from foe is of little importance in TBM encounters. However, because a large number of objects can follow essentially parallel exo-atmospheric trajectories, discrimination of the RV from incidental debris or deliberate decoys can be a significant problem.

The area of coverage of a TBMD system can be obtained by a time-line analysis of the events along the trajectory and of when the defense functions of detection, identification, and interceptor launch and intercept can be carried out. The defended area of coverage is determined by how far the interceptor can fly in the time between interceptor launch and intercept. These times are determined by how well the target must be located and identified by interceptor kinematics and by the last point on the TBM trajectory at which a successful interception can be accomplished. These parameters differ for different threats, different radars, and different interceptors. Some of the issues affecting radar design and some candidate radars for TBMD systems are discussed below. Each of the radar functions is addressed to a top level of detail, including autonomous search, cued search, discrimination, and handover to the interceptor. The performance of each function depends on radar parameters such as power, aperture, frequency, and bandwidth.

The range at which a radar can do autonomous search varies as the fourth root of the power-aperture product times the RCS of the target. Radar power and aperture are limited by cost and transportability requirements. Except to the extent that target RCS may be a function of frequency, autonomous search performance is not a function of frequency. However, since power and aperture are generally cheaper at low frequencies, search radars tend to be designed to operate at lower frequencies as appropriate to U.S. radars.

The threat generally consists of a number of objects—reentry vehicle (RV), debris, deliberate decoys, discarded booster stages, and so on—each of which has an RCS value. Thus, each will be detected at a different range. The booster will probably be the first thing to be detected by a surface radar. It may be detected either in autonomous search or in directed search as a result of cueing by an up-range radar or other detection sensor. After booster burnout, radar data can provide the basis for a good estimate of the booster impact point. For most TBM systems, the booster impact will be close to the RV impact point. Depending on the specific TBM system, the RV may be known to stay relatively close to the booster. This information permits the launch of an interceptor toward the predicted location of the booster. When the interceptor approaches the booster-RV pair, the location of the RV will be resolved by either the surface radar or the interceptor seeker in time to divert the interceptor and kill the RV.

The radar can also do a local search in the vicinity of the booster looking for smaller targets. Since the radar energy can be concentrated in a smaller region, the detection range for these smaller targets can be much greater (often by a factor of 2 or more) than that with autonomous search. Since the radar beam width is narrower at higher frequencies (for fixed aperture size), cued search is generally more effective at high frequency. This is the case for discrimination also. Exo-atmospheric discrimination of both incidental debris and deliberate countermeasures generally relies on looking at the time history of the target RCS or a range- and/or Doppler-resolved RCS map of the target. This requires enough resolution so that different parts of the target appear in different range or Doppler resolution cells. Such resolution is available only at frequencies of S-, C-, or X-band, with the finest resolution at X-band. The use of higher frequencies is limited by attenuation in heavy rain or dense clouds if the radar is oriented toward the horizon and propagates over long distances. For TBMD systems, the radar is generally oriented to search high angles of elevation. In such circumstances the distance that the beam propagates through moisture-laden regions is relatively short. Thus, rain attenuation in TBMD radars may be tolerable at X-band frequencies but not at higher frequencies. For long-range AAW, the radar is designed to search at low elevation angles, and X-band suffers too much attenuation for practical designs. That is why radars with a long-range AAW mission, such as SPY-1 or Patriot, operate at S- or C-band frequencies.

The final function of the surface radar is to hand over the identified target (or threat volume) to the missile seeker. There is a premium for making this handover as accurately as possible for two distinct reasons. First, the requirements on seeker acquisition range are a strong function of the handover accuracy, as is discussed in the section on weapons, below. Second, even if the radar can identify the RV uniquely, if there are other nearby targets, the seeker may not be able to discriminate the RV from another target because its resolution cell has a shape different from that of the radar uncertainty volume.

Ground or aircraft radars generally measure range very accurately and measure angle fairly crudely. The radar uncertainty volume is pancake-shaped, with the diameter of the pancake generally hundreds of times larger than the thickness. A ground or airborne radar forms one of these pancakes for each target in the vicinity of the RV and can pass this information—called a target object map (TOM)—to the interceptor. When the IR sensor on the interceptor looks at these pancakes, it can distinguish different angular positions but does not measure range. Unless the interceptor sensor has a radar capability in addition to an IR capability, the uncertainty region of an interceptor's IR sensor will be conical. If the seeker cone for a particular target cuts through more than one radar pancake, the seeker may not be able to uniquely associate the targets it detects within its cone of uncertainty with one of the radar targets and may, as a result, home on the wrong object. The performance of this function depends on the spacing of threat objects relative to the radar pancake diameter. The radar uncertainty volume is a strong function of radar antenna design and frequency, with higher frequency radars providing narrower beams and higher signal-to-noise ratios, resulting in much more accurate handovers. This handover to the interceptor is an essential fire control function, and the critical need for accuracy is the reason that fire control radars are generally at the highest frequency to propagate in all kinds of weather.

A number of different radars (and other sensors) have been considered for use in ship-based TBMD systems. The capabilities of the current SPY-1 radar and potential upgrades are assessed first, those of other TBMD radars such as THAAD and Patriot are assessed next, those of National sensors, such as the Defense Satellite Program (DSP) and the SBIRS, are assessed last.

The NAD system is a straightforward upgrade of the Aegis AAW system that incorporates modifications to the interceptor, the radar, and the software to permit attacking ballistic targets late in reentry. It does not require very long range or sophisticated discrimination, and the current SPY-1 is suitable for this job.

The NTW system would enable a completely new mission. To get the large coverage, the NTW interceptor (SM-3) is designed to intercept exo-atmospherically, which necessitates launching the interceptor much earlier in the threat missile trajectory. The RV must be detected and identified earlier and at relatively long ranges. The current SPY-1 does not have the sensitivity to detect small RVs at the ranges needed to support the fly-out capability of the SM-3 and must depend on another SPY-equipped ship or other radar to provide track information at longer ranges.

A number of approaches to solving this problem exist, and all of them are being considered. The long-term solution is to develop a new or upgraded radar with sufficient sensitivity. Analysis indicates that such a capability will require an improvement in detection range by at least a factor of 2, which translates into an improvement in sensitivity by a factor of 16 (or 12 dB). The committee was told that the Navy is conducting several "radar roadmap" studies to coordinate radar developments for both NTW and NAD. Some of the candidates for the NTW (Block II) radar include a separate X-band TBMD-only radar similar to the THAAD radar and an S-band or S- and C-band radar to do both TBMD and AAW. The radar detection range depends on the target RCS, and a radar that is adequate for one particular RCS level may be inadequate or overdesigned for a smaller or larger target.

The development and acquisition of a new radar will take a number of years. In the meantime, the NTW system can get some useful capability out of the current SPY-1 radar by taking advantage of ship deployment flexibility in some scenarios and of good knowledge of the threat TBMs in other scenarios. In several important cases (e.g., near the coast of North Korea), the NTW ship could be sited near the TBM launch point and could detect a large RCS booster at relatively short range. It could then do a cued search for the RV before it got out of range. The SM-3 has enough velocity to catch many TBMs even in a near-tail-chase geometry. If the ship cannot get close enough to the launch point to be able to detect the RV, it could use its knowledge of the TBM geometry to launch the interceptor toward the booster and have the seeker acquire the RV in time to divert. However, where the ship must be deployed downrange from the impact point, radar detection of the RV generally occurs too late to conduct a successful intercept. An upgraded radar is required for these cases.

Figures 3.1 and 3.2 show how the requirements for radar range and interceptor velocity can be traded off for both terminal-phase and ascent-phase operation using an exo-atmospheric interceptor.

In this example, the incoming missile's reentry angle is assumed at 45 deg, its altitude at burnout is assumed at 75 km, and its velocity is assumed at 2.5 km/s. This analysis is highly simplified, using flat-earth and straight-line, constant-speed trajectories for both target and interceptor. Although the numerical results are only approximate, the example shows the difference in dependencies on radar range and interceptor velocity between terminal-phase defense and ascent-phase defense.

In terminal defense operation, the forward footprint distance is a measure of coverage (e.g., the distance that the impact point is forward of the defense site). The results show that the coverage can be increased by increasing the radar range to give the interceptor more time to fly out or by increasing the interceptor speed to let it fly further in the same time. The curve for zero footprint corresponds to self-defense.

In ascent-phase operation, the parameter of the curves is the standoff distance, the distance (downrange) from the TBM launch point to the defense site. The curves differ significantly from those for terminal operation. If the interceptor is faster than the target (2.5 km/s in Figure 3.2), a fairly short-range radar may be adequate. It can detect booster burnout, determine the intercept point,

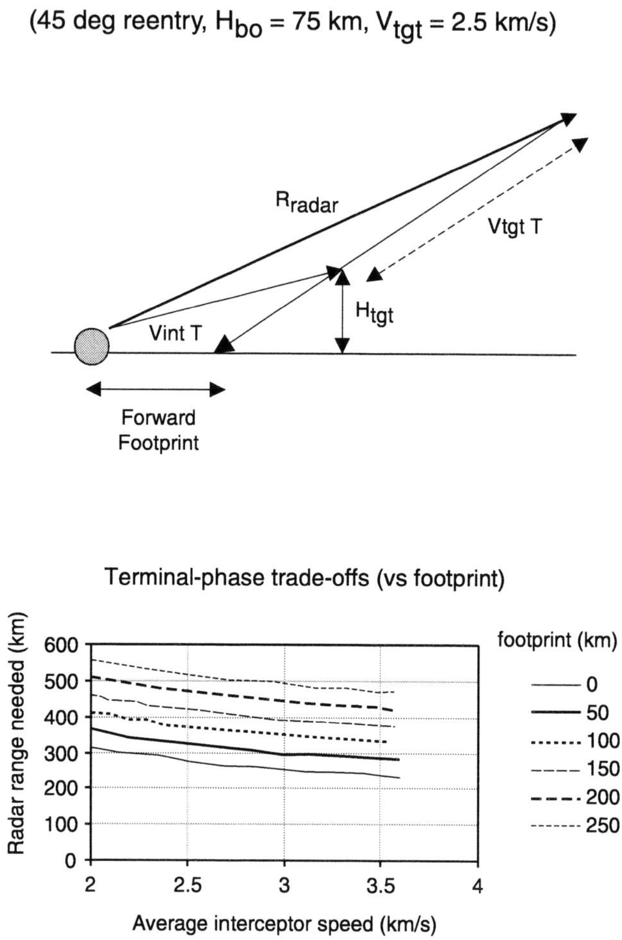


FIGURE 3.1 Trade-offs between radar range and interceptor velocity in a terminal-phase engagement.

and send the interceptor on its way. The intercept point will also be within radar coverage. However, as the interceptor speed decreases, the intercept point gets much further away, and eventually the interceptor cannot catch the target. There is a best location for the defense site—in this case, it is at a standoff of about 150 km (for a 600-km trajectory). If the defense site is closer, the intercept becomes too much of a tail chase, and if the defense site is further away, the interceptor must fly too high to reach the target.

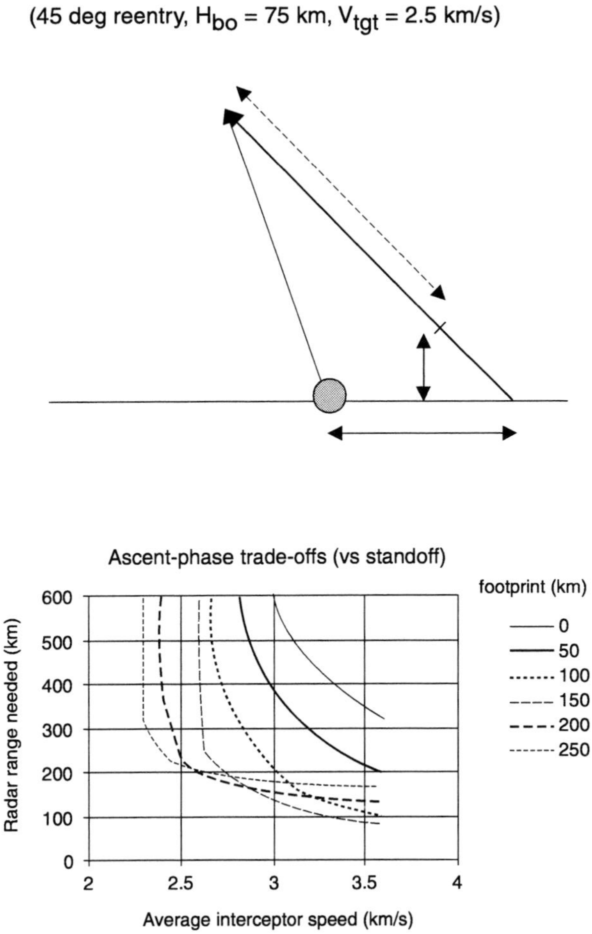


FIGURE 3.2 Trade-offs between radar range and interceptor velocity in an ascent-phase engagement.

A final approach to fire control for NTW is to launch the SM-3 based on data from external sensors. These might include up-range Aegis ships, an up-range land-based radar such as a THAAD, or a space-based system such as SBIRS-low. In some scenarios, these sensors could provide the accuracy needed for fire control, but significant BMC3 changes would be required to exploit this capability. The THAAD radar has the sensitivity comparable to that needed for NTW Block II.

#### 3.2.1.2 National Sensors

Current National sensor capabilities that have relevance to both cruise and ballistic missile defense include the DSP, various signal intelligence (SIGINT) collection programs, and their associated tactical receive applications program (TRAP)/tactical receive equipment (TRE) information dissemination systems. Although these programs provide useful cueing, they are not structured as low-latency systems that can be used to provide direct weapon guidance data.

DSP satellites detect missiles in their ascent phase as soon as they have risen above the cloud bank. If a missile's velocity at burnout is known, or measured, and the direction of flight is provided, then the missile's trajectory and probable intended impact area can be inferred. Unfortunately, the data rate of the DSP sensors is relatively slow. Because it is constrained by the spin rate of the space vehicle, it takes a significant fraction of a minute for the system to declare detection. Since most ballistic missiles reach burnout in less than 2 to 3 minutes, the detection process consumes a considerable fraction of the time available in the ascent stage. DSP is a spin-stabilized spacecraft that uses the spin motion to scan an array of infrared detectors operating in the short-wave IR range to detect the emissions from rocket plumes during the boost phase. The system was designed to have two fixed ground stations (one in the Eastern Hemisphere to detect Soviet intercontinental ballistic missile (ICBM) launches and one in the Western Hemisphere to detect submarine-launched ballistic missiles) and one mobile/deployable ground station. A single DSP satellite gives limited geolocation data relative to the launch site; because of the multisecond frame rate, there is large propagation uncertainty.

If two DSP satellites can view a launch simultaneously (binocular DSP), better geolocation of the launch point will be achieved along with an indication of the azimuth of the missile's trajectory. While such data do not provide precise trajectory information, they certainly limit the volume that must be searched by the defensive radar. This cueing allows the radar to focus its radiated energy into a significantly narrower angular cone and thereby increases the initial detection range of the radar.

DSP is scheduled to be replaced by the SBIRS-high. SBIRS-high is designed to track missiles during powered flight with much greater precision than DSP. This will be possible because the spacecraft will be a three-axis stabilized vehicle. Its optical system is being designed with a large, modern focal plane array. The optical system will incorporate higher sensitivity than is available on DSB. Finally, it is being designed so that it can be adaptively scanned at frame rates that allow detection of much lower intensity rocket plumes. This, in turn, will allow tracking targets at much higher frame rates and will provide a more accurate measure of the trajectory of a missile. In addition, the spacecraft will do much more signal and data processing on board, and the use of cross-linking will eliminate the need for overseas ground entry points. As a result, the system will provide excellent tactical warning and attack assessment and cueing data directly to forces in the field as well as to the National Command Authority.

This system is currently well along in development and the first two spacecraft are scheduled to be delivered within the next 2 years. The current program is planned to deliver a complete constellation on orbit by approximately 2007—the time frame when naval TMD will begin entering the fleet.

The SBIRS-low, while not yet approved for development, is conceptually designed to operate in the visible and long-wavelength infrared (8 to 14 μm) spectra, looking at targets against the cold space background. The task of SBIRS-low would be to provide midcourse tracking of ballistic missiles in flight and hand off the target(s) to a terminal defense system. With its multispectral sensors it could, in principle, provide some discrimination of objects in midcourse.

As a consequence of the decision not to proceed with the future early warning system (FEWS), the SBIRS-high acquisition design parameters were changed to include detection of intermediate-range ballistic missiles (IRBMs) and short-range ballistic missiles (SRBMs). To accomplish a higher scan rate with the greater sensitivity needed to detect and track IRBMs and SRBMs, a two-dimensional, focal-plane array was added along with another detection band at 4.3 μm.

Discoverer II (now a canceled program) was intended in part to be a space-based version of the airborne joint surveillance and target attack radar system (JSTARS). It was designed to function as a ground moving target indication (GMTI) radar, and it would have been capable of providing synthetic aperture radar (SAR) imagery. Various versions of Discoverer II were planned to have between 24 and 57 satellites to provide timely and ubiquitous worldwide coverage.

No space-based analogue of the AWACS AMTI radar that would be appropriate to the problem of ballistic and cruise missile defense exists as a program of record. Although the committee is reasonably confident that the development of such a capability would be technologically feasible within the next 10 to 20 years, questions of affordability exist. Thus, within the next 10 to 20 years, there is relatively little prospect that satellite AMTI radars will become a major contributor to the Navy's missile defense systems.

#### 3.2.1.3 Sensors for Overland Cruise Missile Defense

The concept of Navy platforms providing OCMD requires the ability to detect, track, and intercept cruise missiles that are beyond the line-of-sight horizon of ships at sea. Clearly, an elevated sensor node will be a critical part of any OCMD system. Members of the committee are convinced that both an airborne network link (similar to the CEC) and an air surveillance node are essential to the achievement of a credible OCMD capability. Indeed, the committee believes that Marine Corps expeditionary forces engaged in OMFTS or STOM operations will be seriously hampered if the Navy cannot provide such an elevated node in that kind of threat environment. CEC connectivity through an airborne node appears to be an important part of the Marine Corps' CLAWS point defense concept, along with air surveillance and fire control radar for employing CLAWS.

CEC development itself appears to be proceeding without serious problems. Exercises in Hawaii demonstrated both the concept of operations and the technical achievements required to deploy CEC effectively. The required technology appears to be well in hand, and there do not appear to be any significant technical risks remaining at this point that would delay planned fleet introduction. The primary risks now appear to be inadequate funding for retrofitting the capability on all the firing platforms that could bring missiles to bear on the various air and missile threats to littoral operations.

The prognosis for the development of an airborne sensor node is a different story altogether. The Navy is currently basing its OCMD concept of operations on an upgraded E-2C to provide the airborne node. The E-2C is the Navy's 1960s-vintage solution to its air surveillance requirements for blue-water operations. Its air-surveillance radar has been improved over the years. The current APS-145 radar, which incorporates space-time adaptive processing (STAP), has more near-land and overland clutter rejection capability than the earlier APS-138 radar. However, the overland performance of the APS-145 radar is not adequate to meet the airborne air surveillance requirements for OCMD. The Navy's proposed solution is a hybrid version of the APS-145, which would employ a rotating, electronically scanned array called the ADS-18. Although it is designed to be placed on an E-2C, there is no inherent reason why it cannot be placed on another aircraft such as a militarized 737, C-17, or P-3. Someday a sea-based UAV may be a reasonable candidate for such a radar.

The technology appears to be in hand to develop such a radar, as well as an alternative nonrotating, scanned array with 360-deg coverage that would fit within the envelope of the existing APS-145 radar TRAC-A antenna radome. However, the lead time required to develop and field it will be long. The Navy's planned ADS-18 RMP for developing even the hybrid solution appears to be unaffordable for the Navy at this time; currently the program is not funded. Given the critical importance of an airborne air surveillance node for the achievement of a credible littoral TMD capability, the committee believes that this is a major issue that must be addressed. If funding to support the E-2C RMP is not available, alternative approaches to providing an elevated sensor for OCMD should be considered.

The committee also believes that the E-2C platform's suitability for littoral warfare is open to question. In blue-water operations, its positioning relative to combat air patrol (CAP) fighters and missile-equipped ships can afford it good protection. However, in littoral operations, where the objective is to provide air surveillance for TMD support of troops ashore, it would be more vulnerable to attack by hostile surface-to-air missiles. Operational commanders would need to choose between a standoff distance for necessary survivability and overland radar coverage. During blue-water operations, missile-equipped ships can be positioned to provide continuous air surveillance to cover the battle group when the E-2C is not airborne. In littoral operations, once forces are ashore, the provision of airborne CEC and radar surveillance coverage becomes a much more stressing 24 hours per day, 7 days a week requirement.

Other tactical considerations come into play as well. Owing to its size, which is driven by shipboard launch and recovery constraints, the E-2C has space for only three system operators to do a significant portion of the job done on an AWACS with a crew of about 21. Automation currently offers E-2C system operators some workload relief. However, they are required to work much closer to the overload point in a much more fatiguing environment. This and the more complex, intensive, and dynamic nature of littoral air warfare and missile defense will push operators closer to, if not beyond, their workload limits. The immediate effects on the campaign of this, or the loss of the onboard CEC or radar system, or the platform itself, for whatever reason, must be taken into account, especially with troops ashore. Also, the optimal tactical positioning of the air node for CEC and for air surveillance will probably be different.

Operational availability of the E-2C platform and its transit time to and from station, as a percentage of its total mission time, are other considerations that must be taken into account. The E-2C has 4-hour legs, and the Navy's E-2Cs currently do not have an air-refueling capability. These considerations must be factored into the equation for the number of CEC-equipped RMP E-2Cs that will have to be procured to support the Department of the Navy's TMD concept of operations.

Taken together, these considerations suggest that the Navy's air node platform solution needs to be rethought. One approach might be a CONOPS that routinely depends on the use of the Air Force's AWACS as the elevated AMTI sensor. Another approach might be to employ a sea-based version of the Army's JLENS. The Army claims that the JLENS lift platform can stay aloft in 150-knot winds. If so, the platform can be towed by a surface ship. Unfortunately, the footprint of the JLENS is sufficiently large that it would probably require a dedicated hull for its deployment. Alternatively, emerging long-endurance, fixed- and rotary-wing UAV concepts may offer more affordable and cost-effective air node solutions for CEC and lower-risk solutions for airborne air surveillance in littoral warfare than the E-2C.

Although AWACS, JLENS, and the proposed new E-2C radar would all provide excellent AMTI capabilities against current overland cruise missile threats, their future performance might be degraded by two factors. The first of these would be the introduction of low-RCS cruise missiles. There are limitations in the power aperture gains that can be achieved by such radars. Unless new approaches are adopted, the advantage will eventually shift to cruise missiles. The other problem is that to ensure the survival of an elevated radiating platform, it will generally be necessary to position it at significant distances from the local area of conflict. Since radar performance degrades as the fourth power of the range, the performance of elevated sensors will degrade as their safety is assured by keeping them remote from the area of conflict.

One approach might be to make the elevated node function as a multistatic rather than a monostatic radar. Future variants of the new E-2C radar, JLENS, or the AWACS might function as transmitters in a multistatic system. A multiplicity of UAVs might serve as receivers. Since the UAV receivers might operate relatively close to the area of conflict, there would be significant recovery of propagation loss. This would allow targets with smaller cross sections to be detected. The committee recognizes that operating such a system would be significantly more complex than operating a conventional monostatic radar. However, multistatic radars can detect target glints and can exploit the fact that low-observable cruise missiles are not low-observable from all viewing angles. Thus, multistatic radars offer the possibility of countering further reductions in the nose-on RCS values of cruise missiles.

Finding a more affordable and cost-effective UAV alternative will require rethinking the radar solution. As an example, repartitioning the radar system into a ship-based package and an airborne package through the application of new technologies would reduce the size, weight, and cost. This lower cost would make larger numbers of UAVs more affordable and would provide the operational commander with more reserve capability and lessen the impact of combat losses on the campaign. Since the Navy has not yet funded the E-2C RMP, an examination of other options for the elevated sensor node appears to be both timely and appropriate. If both the E-2C and UAV options turn out to be infeasible, then the Navy and Marine Corps should consider developing a joint CONOPS with the Air Force that is based on the routine use of AWACS to ensure the feasibility of providing OCMD for Marine Corps expeditionary forces ashore.

Based on the foregoing considerations, the committee has concluded that the Navy's inability to fund the introduction of the new AMTI radar with the ADS-18 antenna into the E-2C creates a critical deficiency in its approach to the development of a credible OCMD capability by 2015. The committee also believes that the Department of the Navy may have better options than the E-2C for an airborne CEC and radar surveillance node for littoral warfare.

Although an elevated AMTI radar would be the sensor of choice for OCMD, the committee recognizes that such a sensor may not be affordable. The committee also realizes that although a ground-based radar (GBR) is not an optimized sensor for OCMD operations, a GBR can provide significant radar surveillance capabilities. The Marine Corps actually owns eight TPS-59 radars, which are excellent GBRs that could make a major contribution to OCMD.

Unfortunately, the Marines believe that the TPS-59 is too large for tactical deployments because it consumes too much valuable space and volume on current amphibious ships. Thus, the Marine Corps concept is that the TPS-59 will be flown into theater once an airfield that will accommodate C-5 or C-17 aircraft has been secured. The committee believes there could be scenarios in which no airfield is available in the area of operations that could accommodate the arrival of a TPS-59.

The near-term lack of any kind of effective OCMD sensor should dictate an examination of alternatives. The recent creation of Marine Corps expeditionary brigades (MEBs), which are designed in part to marry up with one of the three MPF squadrons that are forward-deployed around the world, provide such an alternative. The MPF ships have the capability to transport a TPS-59 radar and its associated equipment and can be routinely moved toward developing areas of crisis. The MEB's air combat element (ACE), which includes air defense units, could marry up with the TPS-59 and the MPF squadron in the area of operations for immediate employment ashore.

If the Marine Corps were to place one TPS-59 in two of the three MPF squadrons, then a TBM/OCMD sensor could be moved ashore early in a deployment if required, and it would provide good capabilities for both the Navy units afloat and the Marine Corps units ashore. While the TPS-59 is not a fully satisfactory alternative to an elevated AMTI radar in OCMD engagements, it is significantly better than no OCMD sensor at all. Although the committee acknowledges the limitations of a TPS-59 as an OCMD sensor, it points out that when used in a TBMD role, its performance should be quite credible.

### 3.2.1.4 Sensors for Antiship Cruise Missile Defense

As discussed in Chapter 2, the arsenal of antiship cruise missiles in the hands of potential adversaries is a formidable one. Many of the missiles have characteristics designed to delay detection by shipboard sensors. The principal threat characteristics of these threat cruise missiles are very-low-altitude (sea-skimming) flight, high speed, and reduced radar and IR signatures. Search radars on most Navy ship classes today cannot detect the most serious of such threats in time for missile engagement.

Although a number of improvements in the performance of surface radars have been proposed, ultimately all of them can be negated by some combination of missile speed, low RCS, and low-altitude trajectories. As is the case for OCMD, success in ASCMD will ultimately depend on the availability of an elevated sensor that can provide surveillance and precision track for fire control in regions that are below the radar horizon of ship-mounted radars. The design of such elevated sensors must be sufficiently robust that they will be capable of detecting low-RCS targets at long enough ranges to provide a depth of fire that supports a shoot-look-shoot strategy.

Aircraft carriers and amphibious ships use the SPS-48 (S-band frequency, three-dimensional measurement, rotating antenna) and/or the SPS-49 (L-band, two-dimensional, rotating), which have difficulty detecting targets at low elevation angles. The same is true for the Mk 23 TAS search radars carried on DD-963-class destroyers. Even Aegis's SPY-1 (S-band, three-dimensional, phased-array) radar is taxed against such threats. Sea-skimming flight raises the issue of anomalous electromagnetic propagation due to temperature and humidity variation near the sea surface and also can cause strong multipath fades. These environmental effects can greatly compound the detection problem. Another difficulty results from land background clutter that is inherent in operation in the littorals. Aegis's SPY-1 radar was developed for use in open ocean and is not as effective when operated in the presence of clutter from land background.

The Navy is developing four systems that together will bring a dramatic improvement in detection performance against antiship cruise missiles. The first is MFR, which is planned to provide a horizon search capability greatly exceeding that of any other radar in the fleet today. The MFR is planned for a number of future ship classes, including future aircraft carriers, DD-21, and an LPD-17 upgrade. The MFR (X-band, three-dimensional, phased-array) will not suffer as much as lower frequency (e.g., S-band) radars do from multipath and anomalous propagation effects. MFR will also have good performance near land as well as in open ocean. As an interim measure, to improve low elevation detection capability, the Navy is introducing the SPQ-9B (X-band, two-dimensional, rotating) on some near-term new production ships (carriers and Aegis ships and the LPD-17). It intends to backfit the SPQ-9B on some amphibious ships. At present, there is no plan to switch Aegis new production ships over to the MFR. The committee believes the Navy should reexamine this decision.

The second system that will markedly improve the Navy's capability to detect airborne threats is the SPY-1D(V) radar upgrade on Aegis ships. The upgrade will increase transmitter power, reduce transmitter noise, and possess a number of features to improve its capability against a land background. Signal processing improvements introduced with the SPY-1D(V) upgrade can be backfitted into older versions of SPY-1B and SPY-1D and significantly improve their performance. In self-defense against high-speed, low-observable threats, a cue from CEC (discussed below) or the SPQ-9B can enable the SPY-1D(V) (or SPY-B/D with signal processing upgrade) to detect at longer range and permit weapon employment in some cases where, absent a cue, no shot would be possible.

A third system that will improve detection of airborne threats is a new volume search radar (VSR) to replace the SPS-48 and SPS-49. VSR is being developed primarily for better reliability and economy. It will also provide improved performance against airborne threats above the horizon. Like the MFR, it is planned for future carriers and for the LPD-17 and DD-21 ship classes.

The fourth system that will enhance detection performance against airborne threats is CEC, whose composite tracking capability will take advantage of geometric and frequency diversity to detect and track low-observable vehicles. In CEC, a track is initiated when an individual ship detects an object on multiple radar scans. (This is conventional, except that more than one of the individual ship's radars can contribute to the multiple detections.) The ship initiating the track puts it on the CEC net, and all participants in the net then know the track. Thereafter, any detection by any participant, even if short in duration, is associated with that track. Tests demonstrate that the ability to view a low-observable target from various angles with radars of different frequencies adds significantly to the robustness of the track.

From time to time, infrared search-and-track systems have been demonstrated and proposed for installation on surface ships, especially to enhance low-elevation detection, but they have never reached operational status, primarily owing to their weather limitations.

It is worthwhile at this point to make some observations about the relative importance of area defense as opposed to self (point) defense. First, the SPY-1D(V) upgrade scheduled for introduction with Aegis baseline 7 in 2003 will significantly improve Aegis's area coverage against many threats. Second, cues from CEC can provide additional benefit. In the final analysis, however, even with SPY-1D(V) and cues from CEC, Aegis ships are reduced to near point-defense capability against some low-altitude threats. Therefore detection improvements planned for ship classes other than Aegis are important for their survivability in-theater.

### 3.2.1.5 Sensors for Air-to-Air Combat

For more than half a century, carrier-based fighter aircraft have provided the outer layer of a battle force's defense in depth and escorted aircraft penetrating inland. The current and projected future naval mission—to operate in the littorals and influence events ashore—requires air superiority. Although the technical capabilities of our potential adversaries' tactical aircraft and air-to-air missiles are improving, largely because of Russia's marketing efforts, with few exceptions their air forces are small, and a direct, large-scale confrontation with them is unlikely. However, an air-to-air engagement at the beginning of a conflict could be a logical part of a weaker adversary's response to our naval presence. Early in a conflict an enemy aircraft may have an opportunity to defeat a U.S. Navy or Marine Corps fighter because of asymmetric rules of engagement. The adversary may have permission to fire at will while our fighters are constrained to fire only when positive identification as hostile has been established. Although radars on U.S. naval fighters have greater detection range than those on adversary aircraft and our missiles and our stealth may add to this detection range advantage, freedom to fire at will may enable the adversary to shoot first. Given the political reality that U.S. fighters will remain under strict constraints to limit fratricide and collateral damage, the driving requirement is to achieve combat identification at long range. This is a difficult problem that will probably be solved only by a combination of approaches, one of which is improvements to aircraft radars to allow RF imaging. Some others are track-from-base, SIGINT, and good tracking and data fusion.

E-2C aircraft have for decades provided wide-area air surveillance for carrier battle groups. They continue to be reasonably effective in this role, but the aircraft is looked on to carry out new missions in the future, as discussed in Section 3.2.13, "Sensors for Overland Cruise Missile Defense."

Vectored toward an enemy aircraft by the E-2C, an F-14 aircraft can detect (but not identify) the adversary at long range with its powerful AWG-9 radar. The F-18, including the new E/F versions currently becoming operational, carries a lesser radar. The airborne electronically scanned array (AESA) radar upgrade (APG-79) planned as a P3I program for the F/A-18E/F will significantly improve aircraft capabilities, especially against multiple targets, in response to cueing and through RCS reduction.

#### 3.2.1.6 Electronic Warfare

The surface Navy's capabilities in EW have detection, control, and engagement components, but for convenience, all aspects of EW are summarized in this subsection.

The Navy's principal EW detection asset is the SLQ-32, which exists on virtually every combatant ship in one of five different variants. The SLQ-32 detects the RF emissions of incoming active, radar-guided threat missiles. The Navy had planned to replace the aging SLQ-32 with the much more sensitive and more precise advanced integrated electronic warfare system (AIEWS), but at this writing, the AIEWS program appears to be in jeopardy. Some versions of SLQ-32 have a jamming mode, as will the planned Increment II of AIEWS.

Control of the SLQ-32 is accomplished as an integral part of both Aegis and the SSDS (see subsection "TBMD Weapons," below). Most Navy combatants also have systems to dispense RF chaff and IR flares. Some new ship classes are expected to have the NULKA decoy system.

### 3.2.2 Weapons

As is the case for sensors, the requirements and technologies for weapons in different parts of the TMD realm are diverse. The following assessment therefore is divided into discussions of weapons for (1) TBMD, (2) OCMD, and (3) ASCMD.

#### 3.2.2.1 TBMD Weapons

Ballistic missiles may be attacked with defensive missiles or, under certain circumstances, with directed-energy weapons. At the time of this study, the

Navy had no acquisition programs for lasers or directed-energy weapons. Early concepts for BMD weapons envisaged the use of nuclear warheads on the defensive missile. For a wide variety of well-founded reasons, all such concepts have been abandoned by the United States, and with the exception of laser weapons, contemporary BMD weapons are designed as kinetic-energy weapons (KEWs) or use nonnuclear warheads to negate the target. BMD weapons today engage their targets outside the atmosphere, very high within the atmosphere, or deep within the atmosphere with miss distances small enough to allow fragmentation warheads to achieve lethal damage. Simply put, the problem in using a HTK weapon is to hit a bullet with a bullet. This problem is very difficult, with miss distance requirements less than the radius of the interceptor.

To achieve kinetic kill or the very small miss distances necessary, the weapons in a TBMD system must fly out and acquire, identify, home on an incoming RV coming close enough to fuze (or, in the case of kinetic kill, to collide with the RV), and have high warhead effectiveness. This section focuses on IR homing interceptors such as the SM-2 Block IVA and the SM-3 Blocks I and II. The SM-2 Block IVA is designed to intercept both lower atmosphere air-breathing threats and shorter-range tactical ballistic missiles. It engages tactical ballistic missiles in a deep reentry regime and employs a mid-wavelength infrared (MWIR) seeker, aerodynamic maneuverability, and a fragmentation warhead that is designed to be effective against aircraft and RVs. Because it must engage at low altitude, the SM-2 by necessity has a relatively small defense coverage. The SM-3 operates exo-atmospherically, offering, in principle, a very large defended area. It has the ability to destroy targets before they can maneuver significantly. The price for this capability is the need to deal with lightweight countermeasures that the atmosphere filters out for the lower-tier SM-2. SM-3 employs a kill stage with a long-wavelength infrared (LWIR) seeker, uses thrusters to maneuver, and makes use of the large kinetic energy of a direct hit to achieve a kill.

As discussed in the overview section of this chapter (Section 3.1), a surface radar or other sensor must tell the interceptor where to go. The interceptor needs a burnout velocity ($V_{bo}$) sufficient to achieve a collision point on the trajectory of the RV in the available time. There is a trade-off among the interceptor size/weight, the payload weight, and $V_{bo}$. Both the SM-2 and SM-3 must fit in a vertical launch system (VLS) tube and are thus comparable in size. They embody different trades between payload and $V_{bo}$. The SM-2 has a modest $V_{bo}$ but delivers a heavy payload containing the IR seeker, a semiactive radar seeker, and a substantial fragmentation warhead. It operates in the atmosphere and has significant aerodynamic maneuver capability. The SM-3 carries only a small payload consisting of an LWIR seeker together with navigation and divert systems. An extra propulsion stage has been added to give a much higher $V_{bo}$ and some radar-directed divert capability. The terminal stage or kill vehicle of the SM-3 operates exo-atmospherically and its maneuver capability is obtained by the use of thrusters.

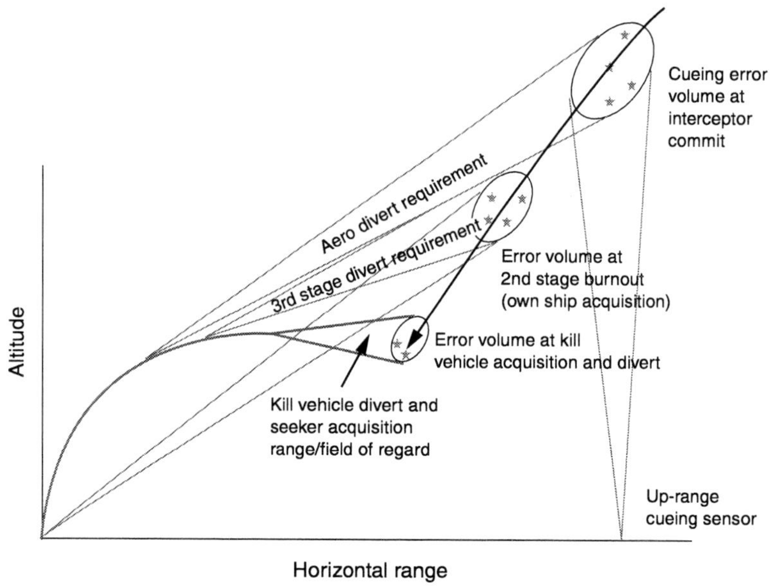


FIGURE 3.3 Kill vehicle look and kinematics constraints versus cueing error volumes (two-dimensional slice)—handover volume effects.

Figure 3.3 illustrates the nature of the progressive handover from cueing sensor to radar to the interceptor. Section 3.1 describes the alternatives and issues involved in detection and handover accuracy, from the early warning sensors to the radar directing the interceptor.

Once the interceptor gets to the vicinity of the target, the seeker must search the radar handover volume, detect all credible targets, identify the RV, and divert toward it. The seeker must have a field of regard and detection range large enough to see an RV anywhere in the handover volume soon enough for the interceptor to divert to and home on the RV with the propulsive or aerodynamic energy available. If the radar can identify the RV from debris or deliberate decoys, the radar's tracking accuracy will determine the handover volume. If the radar cannot identify the RV, the handover volume will be limited by how far the RV could be from a known object such as the booster. If the seeker field of view can cover the handover volume, the detection range is limited only by the seeker sensitivity. If the field of view is smaller than the handover volume, the seeker must scan the volume (for example, using step-stare modes), which will delay detection. The size of the seeker field of view represents a trade-off between the pixel resolution (needed for sensitivity and homing accuracy) and the number of pixels in the focal plane array.

Once the seeker acquires objects in the handover volume, it must decide which one is the RV. Associating its view of the threat with the radar view may do this. If the radar hands over only one object, the seeker looks for the target closest to this point in space. If the radar hands over multiple objects, the seeker tries to match the pattern of objects (TOM). It then homes on the object that the radar identified as the RV. If the radar is not confident of the identification of the RV, or if the threat cloud density precludes confident association, the seeker must do target identification on its own. For a one-color IR seeker, identification might rely on target intensity and scintillation and spatial correlation with handover data. If a two-color IR seeker is available, it will allow inferring the target temperature and emissivity-area product, which provides more powerful discrimination capability. The RV identification process takes some time, which lessens the remaining time (given by the range-to-go divided by the closing velocity) available for divert and homing. The endo-atmospheric interceptor has aerodynamic limits on *g*'s and total divert versus slowdown constraints, and the exo-atmospheric interceptor divert motors have limited *g*'s and total divert velocity. They and the time remaining to closest approach determine how far the RV can be from the center of the handover volume.

After diverting toward the RV, the seeker angle accuracy and the interceptor response time must be such that the resulting miss distance is within the interceptor's lethal radius. For interceptors with warheads, this may be a few meters, but for HTK interceptors, it is a few centimeters. In particular, the HTK lethal radius is smaller than the target, so the interceptor must hit a particular aim point on the target. This is accomplished in the last second when the seeker resolves the target, selects the correct aim point on the target image, and makes a final divert toward that aim point. The feasibility of both of these intercept methods has been demonstrated in BMD research and development programs during the last 20 years.

### *Navy Area TBMD Weapon (SM-2 Block IVA)*

The Navy is modifying its standard missile 2 Block IV into SM-2 Block IVA to provide an improved capability to engage short- to medium-range TBMs. The propulsion stack will be unchanged, and the principal modifications to the SM-2 Block IV are as follows:

- The addition of an imaging IR seeker,
- The addition of an RF sensor to augment terminal fusing,
- Autopilot software modifications to speed the missile's response, and
- A new warhead.

The imaging IR seeker is a focal plane array using indium antimonide (InSb) technology. In the last 10 s of flight, the sapphire dome housing the IR seeker is uncovered. Argon gas is blown over the dome face to separate the aerodynamic shock wave and keep the dome from overheating. The missile operates its semi-active RF seeker for use as a backup if necessary.

While the atmosphere will strip away much of the lighter debris that can surround a TBM warhead, the presence of heavier objects such as the missile propulsion tank or attitude control module can complicate the discrimination task.

The Navy plans to rely on a relatively simple method for handing over the target from the ship radar to the missile. The radar will send to the missile the expected angular position, angular rate, and something akin to the angular acceleration of just one object, the object the ship radar identifies as the target (the part of the TBM complex carrying the warhead). This is in contrast to other U.S. systems, in which the tracking radar transmits a TOM to the missile for handover. The committee was presented data showing that the simple method is proving reliable.

Fuzing is accomplished via an algorithm that combines IR seeker inputs with those from the new adjunct microwave RF ranging sensor. The missile's air target RF fuze is available as a backup.

Among the area TBMD engagement challenges is that of killing a TBM that is "coning," whether inadvertently or deliberately. To inflict sufficient mission-terminating damage, the SM-2 Block IVA be must guided to a location on the target very near its warhead. Therefore target maneuvers of any sort will complicate terminal guidance. SM-2 Block IVA will intercept its target well within the atmosphere, when aerodynamically induced target maneuvers are possible. To counter the helical maneuver effects of coning, SM-2 Block IVA plans to observe the motion of the target, characterize it, and employ a predictive algorithm to estimate the target warhead location at time of intercept.

The SM-2 Block IVA has an explosive warhead designed to be effective in the atmosphere against both air vehicles and missiles.

Taken together, the new elements of SM-2 Block IVA appear to the committee to constitute a moderate development risk, as indeed does the whole Navy area system concept.

### *Navy Theater Wide Weapon (SM-3)*

The NTW system with the SM-3 interceptor is designed to engage exo-atmospherically. Its kill vehicle (KV) seeker cannot operate in the atmosphere and it maneuvers using thrusters that have limited *g* capability. One concern about the SM-3 is its use of a solid-fuel divert and attitude control system (DACS) that is proving difficult to develop. The desire for solid fuel is driven by shipboard safety considerations. The concerns are twofold. First, solid fuel is inherently inefficient in that once turned on it cannot be turned off; coasting can only be accomplished by diverting exhaust gases equally in all directions.

Second, the Navy is using the material rhenium to build the piping necessary to divert the gases. The rhenium must be handled at the high temperatures associated with the solid propellant exhaust and is proving to be a difficult material with which to build reliable plumbing.

By designing the SM-3 only for exo-atmospheric operation, the design can be simpler (and lighter) than the design of an interceptor such as THAAD, which is designed to operate both outside and high within the atmosphere. However, the design also places a number of restrictions on operation in an NTW scenario. Some short-range TBMs, including the SS-21 and Scuds, which have ranges less than 400 km, never get high enough for the SM-3 to engage them. To engage above the atmosphere, intercepts must take place well uprange of the TBM impact point. This significantly limits the coverage in descent-phase engagements, primarily because of the limitations of the radar.

The committee was not presented with detailed analysis of the NTW system except for ascent-phase engagements. Many if not most of the situations the naval forces will face in expeditionary operations will require defense of forces against threats in trajectory descent phase coming from inshore. These engagements are more stressing, as the following discussion illustrates. Figure 3.4

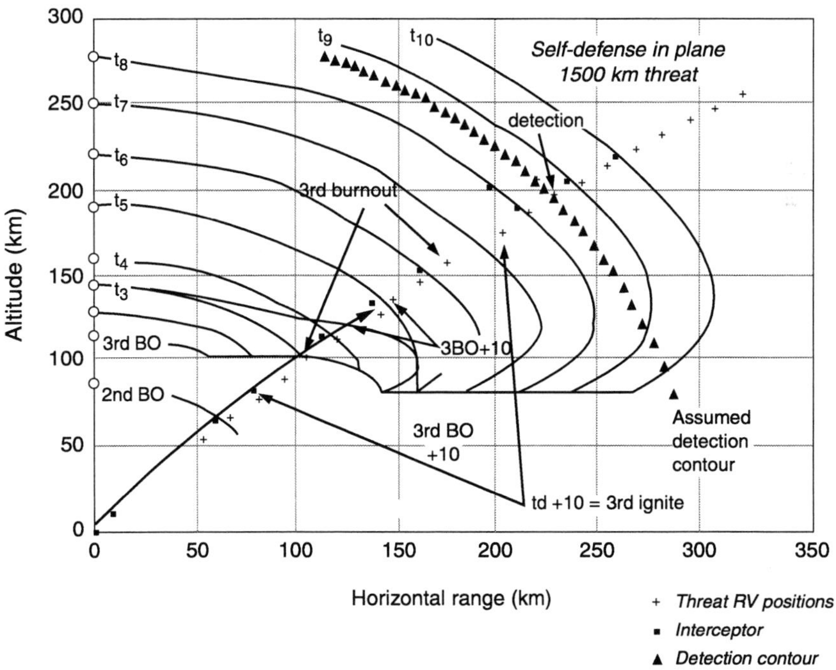


FIGURE 3.4 Self-defense engagement example.

shows a sample graphical engagement analysis for an interceptor with characteristics similar to those of the SM-3 against an intermediate-range TBM. It serves to illustrate the issues arising from the engagement constraints imposed by the various functions that must be fulfilled for a successful intercept. In Figure 3.4, the curved contours represent shortest fly-out time contours for the interceptor, determined by flying out many different trajectory shapes including energy management. The crosses descending from the upper right to the lower left are time ticks along the threat trajectory, and the solid curve from the origin is the interceptor trajectory for the engagement shown.

In this example, it is assumed that a forward sensor like SBIRS-high detects the launch of the threat and tracks it through burnout of the main propulsion. As previously mentioned, the SBIRS frame rate allows determination of azimuth and velocity with sufficient accuracy to project a handover volume forward in time suitable to cue the shipboard radar into a very reduced search volume. It is assumed that the radar begins tracking the target complex when it can see the larger-cross-section booster tank and can commit an interceptor any time after it has the complex in track and has designated the RV. It is assumed here that when the radar has detected and tracked the RV itself for 10 s, it is designated as a target to be engaged. The detection of the RV is assumed to occur at the arc of dots labeled "detection," which for the case shown is arbitrarily chosen to be 300 km from the ship's location at the origin. It is further assumed that the interceptor is launched at the optimum time such that it reaches 80 km (where dynamic pressure is effectively zero) just as the ship's radar designates the RV as a target. This allows the interceptor's third stage to be immediately ignited to divert to the intercept point predicted by the radar track.

During this third-stage divert, the KV seeker is uncapped to begin its search to acquire the RV and other objects in the threat complex. When the third stage burns out, the KV solid DACS is ignited to orient the seeker field of view to the predicted location of the target complex. Assuming a closing velocity of 5 to 6 km/s, three key parameters are as follows:

- The acquisition range of the seeker against the particular RV signature,
- How long the KV needs to search for and acquire the objects and to select the target to engage, and
- The time required to divert and home on the chosen target—a function of the acceleration and total impulse available in the KV.

For the case shown, an arbitrary 10 s was allowed for these functions to occur, as indicated by the contour labeled 3BO +10 s and an arrow that points to where the target is at that time. It can be seen that for this particular set of conditions, the target had not penetrated the boundary, which means that intercept was possible given the stated assumptions.

A careful study of Figure 3.4 shows that:

- The interceptor fly-out capability is more than adequate for descent-phase engagements and the radar detection range for the RV cross sections expected is the first major constraint on a successful intercept,
- The detection range of the seeker for RV signatures to be encountered should be on the order of 100 km, and
- If the action time for the KV to acquire and designate the proper target and to divert and home on the target had exceeded about 12 s, intercept would not have been possible.

Many such cases have been run to establish the defended area and battle space for these assumptions, and several conclusions can be drawn. First, if the interceptor parameters used are representative of the SM-3, the current interceptor fly-out velocity is more than adequate for engaging descending TBMs as far as 50 km forward of the ship provided the radar can detect RVs at the detect range shown. However, the radar limits the battle space, allowing time for only a single shot or salvo against most threats. Second, if the smaller RV cross sections that can be expected in the future cause the detection range to be less than the arbitrary 300 km assumed here, substantial improvements in radar performance will be required just to provide self-protection let alone to project protection ashore. Third, since the intercept is exo-atmospheric and the interceptor cannot take advantage of atmospheric drag to help sort out the RV from the lighter objects, it must rely on more sophisticated measurements of the thermal, spatial, and temporal aspects of the optical signatures of objects in the threat complex. While radar improvements will also help provide more time for kill-vehicle on-board discrimination and homing functions, it will be necessary to increase seeker performance as well to take advantage of that additional engagement time.

Other TMD interceptors, such as THAAD and Patriot PAC- 3, are optimized for specific unique requirements and are generally less suitable for Navy applications. Both of these interceptors are much smaller than the SM-2 and SM-3 because they need to be transportable by air or ground vehicles. THAAD uses a liquid DACS that might support a backup option for SM-3. The PAC-3 active RF seeker may offer some additional robustness for certain kill-stage applications.

The committee emphasizes that at this time, the NTW effort is funded only as a demonstration, not an acquisition program. As a consequence, detailed planning and design are absent. The paucity of realistic engagement data offered for NTW suggests that an inadequate level of systems analysis has been done up to this point in the program. As a result, the committee relied largely on its own analysis to assess the capabilities of the NTW system.

Informal plans presented to the committee project a phase-1 system that barely meets initial requirements. There are strong reasons for proceeding, but only if both the Navy and DOD are committed to follow-on developments that will enable the NTW system capabilities to keep pace with the threat.

### *The Implications of Countermeasures*

There are three important aspects to exo-atmospheric discrimination ability for systems having a NTW-like architecture. First, even in the absence of penetration aids and even if the separated reentry vehicle has been designated by a radar, the complex of objects that accompany the warheads, described in Chapter 2, requires an interceptor to distinguish among several objects at the time it searches to acquire the designated target. The ability to do this successfully has been demonstrated several times with two-color sensors possessing appropriate onboard processing. The keys to the success of this process are (1) sufficient observation time and (2) the closing geometry of the interceptor with the target, which provides important range-dependent information not available to any other sensor. This information and the need for observation time place a high premium on the interceptor sensor acquisition range as well as the interceptor commit time. Both parameters become even more important when countermeasures such as chaff and decoys are introduced.

The use of sophisticated countermeasures implies complexity of design in threat TBMs. Complexity, in turn, generally increases both weapon cost and probability of failure. To date, adversaries that have launched TBMs at U.S. forces or those of its allies generally have been more concerned about successful delivery of a weapon on a target than about penetration aids. The committee believes there is a high probability that the countermeasures that might be employed by future adversaries will be simple responsive measures that embody low technology and low risk for their implementation. These might include tactics such as tethering a simple, lightweight radar decoy to the rear of a separating RV. Such a decoy would be aimed at drawing the HTK vehicle off the target during the last seconds, when it becomes resolved as a separate object.

The committee did not have an opportunity to undertake an extended analysis of the effect of countermeasures on the aim point selection problem and the miss distance, which create different problems for exo-atmospheric and endo-atmospheric intercepts. While a low-altitude interceptor must deal with planned or inadvertent aerodynamic maneuvering such as that seen on Scuds during the Gulf War, intercepts outside the atmosphere must deal with tethered objects or appendages that create unusual dynamics as well as with modification of the optical signature. In both cases, studies have shown that predictive guidance filter algorithms will work provided there is sufficient observation time to determine the pattern of the dynamics.

Another approach is to increase the lethal radius of the kinetic-kill vehicle. A lightweight means of extending the kinetic-kill effective radius using an inflatable "kill enhancement device" was demonstrated in the exo-atmospheric reentry vehicle interceptor system (ERIS) program in the early 1990s.

In any case, there needs to be a response to these countermeasures as they emerge. If this response is provided for with prudent R&D efforts backing the initial system, the committee believes that defense effectiveness can be enhanced in an evolutionary manner.

### *Growth Paths for a More Robust SM-3 Interceptor*

If SPY radar performance is improved as recommended in this report, the KV acquisition range, discrimination capabilities, and divert capability become the limiting characteristics for the performance of the theater-wide SM-3 interceptor. In addition, if countermeasures such as simple tethered objects are introduced, it may be desirable to add a kill enhancement device similar to the light inflatable device demonstrated on the ERIS program.

It can be seen that each of these performance improvement measures has a kill vehicle weight penalty. For this reason, the committee also endorses the work to develop a 21-in.-diameter, second-stage rocket motor for the SM-3. This will allow retaining the fly-out performance with a heavier KV and also offers some ancillary benefits. That same motor could be used for a strike variant of the standard missile that could be employed for prompt counterfire and other strike missions, allowing useful-size payloads and an extended reach.

### 3.2.2.2 OCMD Weapons

In keeping with its commitments to the concepts of OMFTS and STOM, the Marine Corps has eliminated its Hawk capabilities. The only missile defense weapon now available to Marines ashore is the relatively short-range stinger missile. The stinger was designed to engage low-flying aircraft and helicopters. Its effectiveness against low-flying cruise missiles may be expected to be quite limited.

The Marine Corps is conducting an exploratory development program to develop a lightweight, mobile missile defense system utilizing the AMRAAMs mounted on an HMMWV. The resulting system, called CLAWS, should provide the Marines with reasonable, forward-deployed firepower. In light of the Marine Corps' current need for an OCMD weapon, the committee supports this effort and believes it should be accelerated.

The main problem with the CLAWS concept is that no sensor is available to provide it with beyond-line-of-sight target information. For targets that are within the line of sight, a number of conceptual sensors, such as those listed above, might be employed if (as planned) a CEC is incorporated into the CLAWS development.

Currently the Navy has no weapon that can defend against overland cruise missiles that are below the horizon of ship-based radars. The provision of an outer layer of area defense against overland cruise missiles will require—in addition to a resolution of the issues related to sensors—a sea-launched missile that can be directed by an airborne platform. The development of such a weapon would require a substantial effort. A number of CONOPS and weapon guidance options would need to be explored. Among the CONOPS and terminal guidance options that might be considered are the following:

- A sea-launched semiactive missile that flies to a designated point where it can then home on its target based on the target illumination provided by an airborne illuminator. In this concept, the aircraft that detects and illuminates the cruise missile target calls for a missile launch by an appropriate ship.
- A missile with an active radar or IR seeker that in its midcourse phase could be command-guided by an airborne surveillance platform. When the missile is close enough to detect the target with its own seeker, it transitions to autonomous control for the terminal phase of the engagement. As in the previous option, the surveillance aircraft that detected the target missile would call for a missile launch by an appropriate ship.

Aircraft with AMRAAM missiles may have some capability against low-altitude cruise missiles, but sustainability considerations dictate that most of the defensive coverage for the area be provided by missiles launched from surface ships operating offshore. Except against cruise missiles that can be tracked by those ships, the Navy has no such capability at present, and the committee was unable to identify a program of record to develop such a capability.

### 3.2.2.3 ASCMD Weapons

SM-2 Blocks III and IV today provide Aegis's area AAW engagement capability. SM-2 Block III is a medium-range weapon with a semiactive RF guidance augmented by a nonimaging IR seeker for countermeasure robustness. SM-2 Block IV is a long-range weapon designed to handle fast, high-flying threats. SM-2 Block IVA, now under development for TBMD, has the same propulsion system as (but greater maneuverability than) Block IV. Block IVA's imaging IR seeker will not be used against cruise missile or aircraft threats.

Apart from air defense systems employing standard missiles, the Navy's current engagement systems for hard-kill ship self-defense include the NATO sea sparrow missile system (NSSMS), the RAM, and the phalanx close-in weapon system. These weapons are described in Appendix D.

NSSMS is the principal air defense system on today's aircraft carriers, on amphibious vessels (LHDs), on DD-963 destroyers, and on some under way replenishment ships. It employs the venerable semiactive sea sparrow missile, which has limited capability against today's threats. Sea sparrow has a range of about 10 miles.

RAM, now deployed in its Block 0 version on many amphibious ships and DD-963 class destroyers, is a shorter-range missile. Block 0 RAM is a fire-and-forget missile that homes on an incoming missile's radiation until the RAM's nonimaging IR seeker can acquire it. It is only effective against threats that employ active radar for terminal guidance. The Block I variant of RAM overcomes this limitation. Successfully completing its operational evaluation (OPEVAL) last year, RAM Block I has a much wider field-of-view IR seeker, which can acquire incoming threats based on shipboard radar handover alone. The RAM Block I is planned for installation on carriers and many amphibious ships. RAM's ability to attack at minimum range gives it capability against threats difficult for other missiles: some threats maneuver at a distance from the ship but reduce their maneuvering as they draw close to ensure that they hit the ship. The principal disadvantage of RAM's short range is its inability to handle high raid densities.

The close-in weapon system (CIWS) is on virtually all combatant ships. It is a closed-loop system in which a radar tracks both the threat and a gun's projectiles, judges the distance by which the projectiles are missing the incoming threat, and adjusts the gun's direction of fire. Its very short range makes it a last-ditch defense. CIWS was first introduced 20 years ago. Although many variants and upgrades of CIWS are now operational, there are some threats the system cannot handle. The Navy plans to replace CIWS with RAM.

The Navy is developing the evolved sea sparrow missile (ESSM) to handle emerging cruise missile threats, especially fast and highly maneuverable ones. The ESSM is a greatly improved upgrade of the sea sparrow; it provides a more powerful rocket motor, better aerodynamic control, and a new guidance system. ESSM is currently in development flight test, and at the time of this writing it is having some difficulty with radome failures. Navy presentations to the committee showed that either ESSM or SM-2 Block IVA is necessary to give surface ships adequate self-defense against the most serious air threats expected to emerge in 2005 or so. At present, the Navy plans to install one or the other missile only on Aegis ships, with ESSMs packed four missiles per vertical launch system (VLS) cell. The Navy once planned to install ESSMs on other ship classes, but to do this it needs an ESSM launcher, for which there is no program of record.

The SM-2 Block IVA provides Aegis ships with another weapon that could be used in self-defense, as well as in area defense or TBMD, as discussed elsewhere. Neither SM-2 Block IVA nor ESSM would be adequate against potential future air threats employing certain advanced countermeasures.

The MFR discussed above could be made to serve as an illuminator for semiactive missiles such as the standard missile or ESSM. The MFR will have a phased-array antenna, which should enable it to handle multiple missiles in terminal guidance, thereby improving defense against high raid densities. This is another reason to consider its use on Aegis ships.

The combination of the VSR, MFR, a weapon control architecture similar to that in Aegis, the SSDS, ESSM, and a robust electronic warfare (EW) capability should provide future combatants other than Aegis an adequate self-defense capability against most threats in the near term. Again, the threat can be expected to continue to increase, and future upgrades—to handle, for example, advanced countermeasures—will surely be required.

### *Weapons in Air-to-Air Combat*

Current and potential future carrier-based fighter aircraft include the F-14, the F/A-18 (undergoing significant upgrade to the E and F versions), and the joint strike fighter (now in competitive flight test). The E-2C aircraft provides early warning and fighter direction. Marine air squadrons are often incorporated into the carrier air wing and most likely will continue to be in the future.

The Navy and the Air Force have adopted different approaches to fighter aircraft design. The Air Force's F-22 pushes the state of the art; it is fast, stealthy, and expensive. The F/A-E/F is less expensive and somewhat less capable; it will rely more on jamming support, on networked operations, and on air-to-air weapon effectiveness. Since the principal air-to-air weapons are developed jointly, the different approaches to fighter design can lead to differences of opinion on weapon requirements.

The F-14 can carry the long-range AIM-54 (Phoenix) missile. The active-RF-guided Phoenix was developed for the outer air battle the Navy was prepared to fight during the Cold War. Its inability to achieve long-range combat identification limits its usefulness today.

The AMRAAM is carried on F-14 and F-18 aircraft. It is initially command-guided, with communication via the aircraft's radar, and then employs an active radar for terminal guidance.

The IR-guided AIM-9 sidewinder missile used in short-range air-to-air combat is undergoing a significant upgrade made necessary by the greater maneuverability of new air-to-air threat missiles such as the Russian AA-10, which is being widely exported.

As a last-ditch weapon for air-to-air combat, current Navy fighter and strike aircraft carry a small-caliber cannon that can also be used for air-to-ground strafing. The utility of such a weapon has long been debated (the F-4 was built without one), and today the joint strike fighter operational requirements document (ORD) requires a "missionized" gun, that is, a weapon that can be easily removed and reinstalled.

The Navy appears to be moving toward greater reliance on networked operations in air-to-air combat. Long used to cueing from the E-2C, fighters are now developing capabilities for fighter-to-fighter off-board targeting. Link 16 may be key to this communication.

### *Weapon Control in ASCMD*

The speed of many antiship cruise missiles, their ability to delay detection, and the adversary's potential to coordinate attacks so that greater numbers of attacking missiles arrive in shorter periods of time combine to make reaction time and firepower the principal challenges in shipboard weapon control.

The Navy's answer to the reaction time and firepower challenges is to rely on automation and to provide doctrine allowing the commander to depend on an automated response in high-threat conditions. To provide the capability, the Navy has also had to meet stringent launch control, launcher design, and illuminator requirements to fire and guide semiactive air defense missiles. The first implementation of this was in Aegis. For decades an Aegis ship has been able to have multiple missiles in the air against an incoming threat just a few seconds after establishing a firm track on it.

In recent years, the Navy has implemented an ad hoc capability for automated fast reaction in its other combatant classes. The SWY-1, -2, and -3 weapon control systems in these ships have a reaction time with a RAM that can rival Aegis's with a standard missile. The Navy now plans to replace this ad hoc weapon control capability with the SSDS—a modern, open, distributed architecture founded on a local area network (LAN). SSDS treats sensors and weapons as LAN access units, permitting easier replacement.

The Navy also plans to evolve Aegis toward an open architecture. When this has been done, the Navy will have the opportunity to standardize the command and decision (C&D) element of air defense systems on its various ship classes.

The Navy will soon have three levels of BMC3 systems applicable to area AAW, providing for three levels of operations:

- An individual ship providing air defense to other forces,
- A battle force, and
- Joint forces in a regional theater.

The command and decision systems on Aegis cruisers and destroyers provide the first level; the CEC will enable BMC3 for a battle force; and the system being developed to support the area air defense commander will provide BMC3 for air defense within a theater. These systems are as applicable for TBMD as they are for area AAW and CMD.

The Navy and Marine Corps BMC3 for theater missile defense is discussed in more detail in the next section.

### *Area Antiair Warfare*

SM-2 Blocks III and IV today provide Aegis's area AAW engagement capability. SM-2 Block III is a medium-range weapon with a semiactive RF guidance augmented by a nonimaging IR seeker for countermeasure robustness. SM-2 Block IV is a long-range weapon designed to handle fast, high-flying threats. SM-2 Block IVA, now under development for TBMD, has the same propulsion system as (but greater maneuverability than) Block IV. Block IVA's imaging IR seeker will not be used against cruise missile or aircraft threats.

## 3.3 BATTLE MANAGEMENT COMMAND, CONTROL, AND COMMUNICATIONS

### 3.3.1 Introduction

BMC3 is the overall process and supporting capability for realizing the tactical direction and coordination of sensor and weapon assets. It is considered here in the context of missile defense, but that is taken to include the necessary coordination of offensive and defensive operations. The set of functions carried out in this process is shown in Figure 3.5. A BMC3 system is implemented by allocating these functions to humans or computer processors, with data flowing between the functions over communications links. Since the functions are typically performed at spatially separated locations, the BMC3 system is a distributed information processing system.

As shown in Figure 3.5, sensor data are processed to determine the number and location (tracking), identity (identification/classification), and status (kill/battle damage assessment) of hostile, friendly, and neutral aircraft and missiles. The sensors can be organic naval assets, theater assets or those of other military services, or National ones. These sensor data are consolidated and interpreted (integration of situation inputs) to develop a tactical situation picture to serve as the basis for decision making. The objective obviously is to have a situation picture that is as complete as possible and minimal uncertainty in the location and identification of the objects shown.

Platform and weapon management/control is composed of three separate subfunctions—platform direction, weapon assignment, and fire control. Platform direction pertains to the airspace management and deconfliction of friendly assets in the region of interest, including the coordination of offensive and defensive assets. Weapon assignment designates the particular weapon (or weapons) to attack a given target; it can involve the coordination of assignments across the weapon systems of different military services.

Fire control is the process for guiding the defensive weapon to the target, e.g., guiding a missile from its time of launch until it is able to acquire the target by itself. In a highly distributed configuration, the sensor providing the guiding information, the platform launching the defensive missile, and the node determining the control commands could all be separated by significant distances.

Sensor management/control assigns sensors to support the engagement of current targets and provide data for the detection, tracking, and classification of new targets. Communications management/control allocates communications resources (e.g., links and time slots) to support data transfer between the functions. As shown in Figure 3.5, both sensor and communications management/

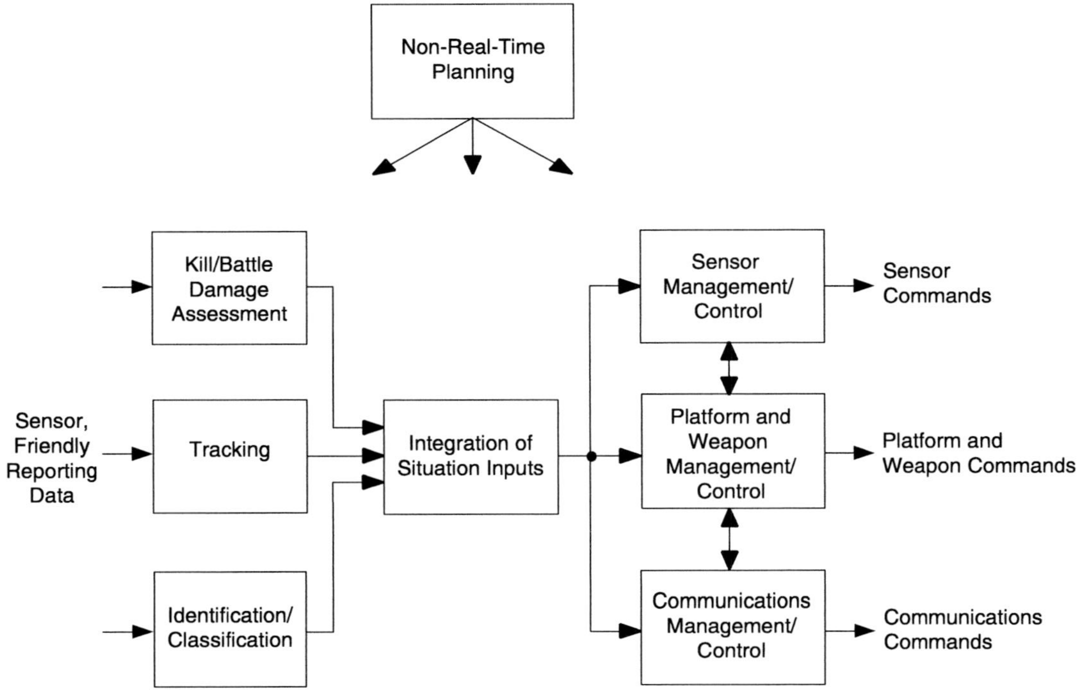


FIGURE 3.5 Missile defense BMC3 functions.

control couple into platform and weapon management/control, as is necessary, for instance, to carry out fire control.

All the real-time functions noted above are performed according to plans developed in non-real-time planning, which may need to be revised in near-real time. For example, non-real-time planning determines initial platform locations, assigns sensor coverage areas, and provides rules of engagement.

The next section elaborates on operational considerations; the sections after that discuss how current and planned systems and programs relate to the achievement of these BMC3 functions.

### 3.3.2 Operational Considerations

#### 3.3.2.1 Need for Flexibility

Looking at recent operational experience from the BMC3 perspective, one fact stands out clearly: The BMC3 concepts and procedures brought into these operations were significantly altered and augmented in the face of the operational realities. In the recent Kosovo operation, for example, when the rule of engagement (ROE) was imposed that all targets had to be observed visually prior to attack, Navy F-14s with their large display screens were used to provide this visual confirmation to attacking Air Force F-16s. This process had not been anticipated prior to the operation. Likewise, when it became necessary to precisely determine the coordinates of the visual imagery taken by UAVs, a method adapted from techniques developed by the intelligence community was used to impose coordinate registration on this imagery.

The basic point is that information exchange and processing will have to be carried out in previously unanticipated ways. This can often involve the exchange of information across Service systems or from National systems. Also, the committee believes that different users will often require different information from the same sources. In fact, no one can really specify a particular user's information needs other than the user.

A similar situation can be expected to pertain in theater missile defense. No matter what one thinks the BMC3 situation will be, it will probably change. Factors involved include the unexpected failure of systems, unanticipated features of the threat, valuable information from new sources, newly imposed ROEs, and so forth. A further dimension is the need to adapt to the rapid evolution of technology.

BMC3 for theater missile defense is currently prescribed in a rather rigid manner with no natural provisions for the operational flexibility and adaptability that are likely to be required, either in terms of operational concepts or the underlying technical capabilities. Link 16 is a good case in point. It requires that all participants (sensors and weapon platforms) to a Link 16 network be spelled out in advance. In fact, several days of advance effort may be required to set up the necessary network configuration. Thus, the dated technology of Link 16 manifests itself by inhibiting operational flexibility.

Of necessity, all operations are likely to be jury-rigged. Operational concepts and the technical underpinnings should allow this ad hoc assembly of components and information exchanges to become a normal process rather than constantly repeated exceptions. Current Internet and Web concepts provide some elements of the solution (although this does not suggest that the Navy/Marine Corps should use the Internet in implementing the solutions). Commercial businesses often use Internet and Web technologies to assemble ad hoc participants and information sources to gain important new business capabilities. One outcome of the further development of this theme is a systems engineering process that accommodates the introduction of unplanned resources and capabilities (Appendix C).

These ideas on the flexible composition of forces are reflected in current naval thinking on network-centric operations.[3] However, although the Navy in general appears to espouse network-centric ideas strongly, such concepts were almost totally lacking in the TMD briefings and reports presented to the committee. As the Department of the Navy and joint community move forward with developing TMD concepts and capabilities, network-centric ideas need to become much more prominent.

### 3.3.2.2 Sample Scenarios

The committee postulates two scenarios that will place its analysis in context and make its observations concrete and easier to understand. Realizing fully that there is a very broad range of potential scenarios, the committee has selected neither the simplest possible scenarios nor those that appear most often on briefing charts, but simply plausible ones that illustrate the points made in this section.

The first scenario involves theater-wide defense against ballistic missiles. It involves a joint commander afloat, where a naval NTW system is augmented by a PAC-3 battery for city defense. The commander has elected to tie the SBIRS-high system into the defense since it can provide very good early cueing of missile launches that are expected to take place far inland. There is reason to believe that the incoming missiles may carry chemical or biological payloads, so the National Command Authority is keenly interested in knowing the current situation as it unfolds in real time and perhaps also in giving some detailed guidance to the commander in the field.

In this first scenario (Figure 3.6), the principal BMC3 challenges are to weave together, as far as is feasible, the data from the Navy's SPY-1 radar, from

[3]For additional reading on network-centric operations, see Naval Studies Board, National Research Council. 2000. *Network-Centric Naval Forces: A Transition Strategy for Enhancing Operational Capabilities,* National Academy Press, Washington, D.C.

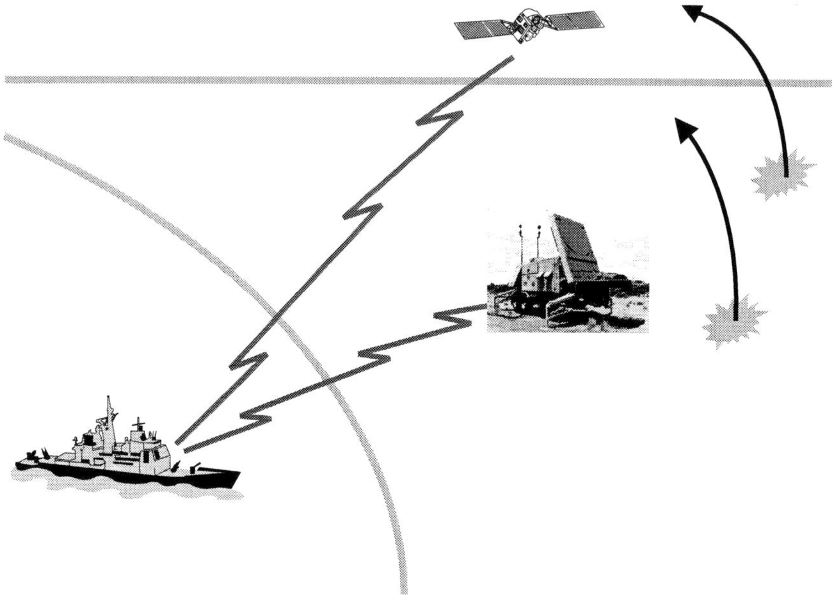

FIGURE 3.6 A theater-wide ballistic missile defense scenario.

the PAC-3 radar, and from the SBIRS sensors. The data will be used to decide whether the NTW system or PAC-3 will attempt to intercept a given incoming missile. The decision might depend on which has the better shot and where the debris may land. An additional requirement might be to keep the National Command Authority fully informed and in the decision loop as attacks unfold.

The second scenario (Figure 3.7) involves cruise missile defense. Here it is assumed that the fleet has a dual defense role: it must protect itself and extend protection to Marines who have maneuvered far inland. In this example, the purely naval force has been augmented with an AWACS presence that serves to detect cruise missiles while they are still far inland. Some of the cruise missiles may contain explosive warheads (e.g., for attacking the Navy's ships) while others may contain chemical or biological payloads aimed at the Marine deployments and nearby cities. The enemy is postulated to have chosen to launch cruise missiles directly through the Navy's aircraft so that friendly planes, commercial aircraft, incoming cruise missiles, and interceptor missiles could all occupy the same airspace.

This second scenario poses extremely complex and difficult challenges for BMC3. First, the ships afloat and the AWACS must share a highly detailed

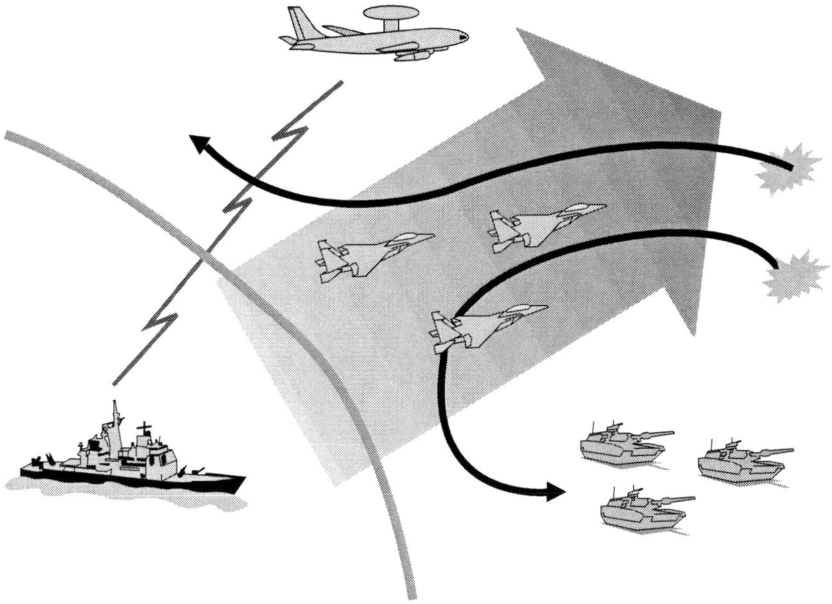

FIGURE 3.7 An overland cruise missile defense scenario.

picture of all the objects in the airspace (e.g., a SIAP) so that they can distinguish enemy missiles from friendly planes and from neutrals. To a large extent, this picture must be synthesized from radar inputs from both the afloat and aloft sensors since the cruise missiles may have relatively low observability. Second, this air picture must somehow be related to the ground picture so that the naval shooters know the current location of the Marine Corps. Third, the BMC3 system must help the commander decide in real time on the best locations for intercepting the cruise missiles so as to minimize collateral damage caused by their falling debris (chemical and biological). Fourth, friendly planes may need to be diverted in real time so that the Navy has a clear shot at the incoming cruise missiles. Finally, the National Command Authority may require an accurate and highly timely picture of the entire battle as it unfolds in order to oversee, and perhaps override, the local commander's decisions.

Note that neither of the scenarios is Navy/Marine Corps-only—one involves forces of the Army and the other forces of the Air Force. As such, they introduce inter-Service complexities greater than those in Navy/Marine Corps-only scenarios. In addition, both scenarios intertwine the National Command Authority fairly tightly into engagements that otherwise must move along tactical time lines. The committee believes that neither of these complexities is unlikely. Indeed, it believes that joint activities, very likely with a direct tie into the National Command Authority, are more likely to occur than a simpler Navy/Marine Corps-only scenario. In such situations, a joint commander is likely to ask for as much help as is possible and is likely to be given whatever is feasible.

### 3.3.2.3 The Importance of BMC3

For the most basic threat scenarios, the Department of the Navy's development priorities could plausibly be ordered as follows: missiles first, then radar, and last (and almost least) BMC3. That is, for the simplest threats, the BMC3 components are relatively easy to envision and implement and can safely be assigned a far lower programmatic priority than the development of good missiles and radar upgrades. In the more stressing threat scenarios, however, the BMC3 component begins to move to the forefront and becomes relatively more and more important.

Thus one can say broadly that plans for wide-area systems—against both ballistic and cruise missiles—place a higher premium on the BMC3 system than do plans for more local defense systems (the NAD system and ship self-defense). This is because wide-area systems require a higher degree of coordination between geographically dispersed platforms. The sensors may be quite far away from the shooters in wide-area systems. With local-area systems, by contrast, the sensors and shooters can be collocated on a single platform. Simply put, wide-area systems begin to demand network-centric solutions, while for the less stressing local threats, classic platform-centric approaches will suffice.

Table 3.1 presents, in highly condensed form, a range of situations that may be encountered. The implications are discussed in the paragraphs that follow.

TABLE 3.1 Implications for BMC3 of Specific Naval Missions

| Mission | Implications for BMC3 |
|---|---|
| Antiship cruise missile defense | Handled acceptably by CEC evolution |
| OCMD | No plausible capability without sensors from external sources (AWACS, UAVs, JLENS) |
| NAD | Appears in good shape but would benefit from external cues (e.g., SBIRS-high) |
| NTW | Limited capability for many missions without external sensors |

For NTW and overland cruise missile defense, BMC3 begins to assume a critical importance. In general, the farther the Navy must stand off from a hostile coastline, the more it will be forced to rely on external sensors. Its reliance on BMC3 systems will grow accordingly.

There is every reason to believe that both ballistic- and cruise-missile threats will grow more stressing over time. Thus, even those scenarios that can at present be managed by platform-centric approaches—namely, area and ship self-defense—will in relatively short order become too stressing for that simple approach. They, too, will begin to require more complex BMC3 solutions. This is not a new phenomenon. The rise of the cruise missile threat led to the relatively complex and distributed CEC system for ship self-protection. In other words, as the threats become more sophisticated and more numerous, distributed BMC3 systems will grow more important. Such systems are, in essence, the glue that binds the widely distributed sensors and shooters that form the protective shield for naval forces.

In summary, BMC3 is already critical for the Navy's more stressing threat scenarios (NTW and overland cruise missile defense). As time passes, it will also become more and more critical for even relatively local types of defense since the evolution of the threats will require ever-more-complex defensive systems.

One further aspect of missile defense bears special mention. Overall, increasing threat levels lead to a radical physical separation of the sensing, shooting, and command components of the entire system—and, indeed, lead very quickly to systems in which these various functions are handled across Services. For example, the Army may provide the radar and the Navy may provide the missiles. Thus, the perhaps inevitable response to ever-growing missile threats leads to a system that is "joint" to a profound degree and as such may require a major change in Service cultures.

The Department of the Navy should therefore be placing a fairly heavy emphasis on its distributed BMC3 architecture and systems (Figure 3.8). These systems are already important for the more stressing naval missions and will rapidly become critically important even for missions that can currently be handled by platform-centric BMC3 systems. These systems and the associated architecture are discussed next.

### 3.3.3 BMC3 System Architecture

Figure 3.9 shows the Navy's BMC3 system architecture based on current and near-term systems.[4] This architecture is largely the result of the historical development of capabilities rather than a top-down system design.

---

[4]Warner, Eugene, "BMC4I/Interoperability for Navy TAMD," briefing to the committee on June 28, 2000, Program Executive Office, Theater Surface Combatants, Arlington, Va.

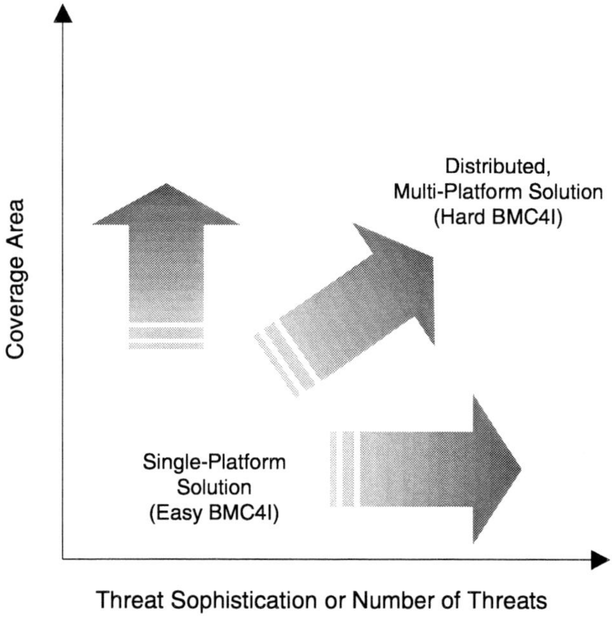


FIGURE 3.8 Increasing area coverage and/or threat level will require distributed BMC3.

The left-hand block in Figure 3.9 refers to time-critical (but not real-time) decision making and non-real-time planning. Various data feeds, including from intelligence sources and communication means, enter the global command and control system-maritime (GCCS-M). Some tactically derived information is input to the left-hand block, but an additional large source is the tactical digital information links (TADILs) shown in the right-hand block. The TADIL inputs are processed in the command and control processor (C2P) for use by the rest of the BMC3 system.

A more rational, modern design for the overall BMC3 system would recognize that there is significant commonality of purpose and use of the data inputs and processing in the right- and left-hand blocks. In particular, a more modern approach would move from the many special-purpose systems shown here to a configuration based on common standards and general-purpose communication and computing capabilities.

The center block in Figure 3.9 represents real-time and near-real-time decisions to allocate and launch defensive ship-based missiles. The components involved are the advanced combat direction system (ACDS), the SSDS, and the Aegis command and decision and display system (C&D/ADS). This module

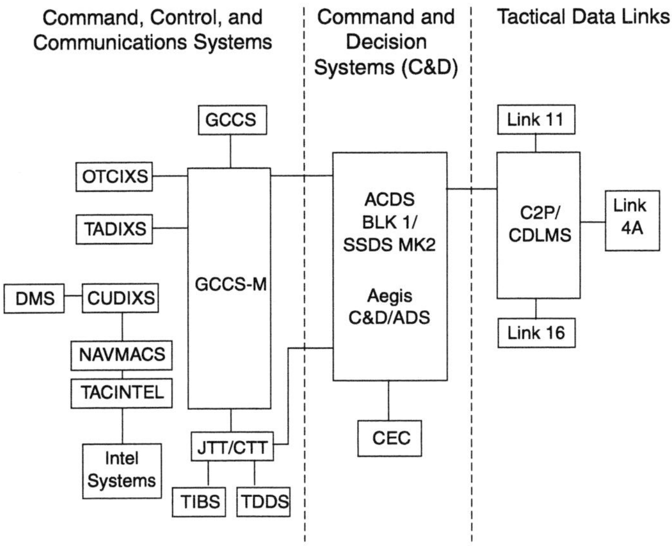


FIGURE 3.9 BMC3 system architecture. DMS, defense message system; OTCIXS, officer in tactical command information exchange subsystem; TADIXS, tactical data information exchange system; CUDIXS, common user data information exchange system; NAVMACS, naval modular automated communications system; TACINTEL, tactical intelligence information exchange system; JTT, joint tactical terminal; CTT, commander's tactical terminal; TIBS, tactical information broadcast service; TDDS, tactical receive applications (TRAP) data dissemination system; ADS, advanced display system; CDLMS, common data link management system.

receives its data input from the CEC, which are netted Aegis radars, as well as from the TADILs and GCCS-M data sources. The Navy has been experiencing interoperability problems as it upgrades the components of this module. Those problems appear to be in the process of resolution. In the longer term, the Navy intends to replace this module with the so-called Aegis common command and decision system (CC&D), which will be based on a modular, open architecture that should help to minimize future interoperability problems.

The TADILs and CEC are essential for providing the sensor input necessary for the BMC3 process. Of the three TADILs shown in Figure 3.9, Link 16 is the primary one in Navy plans. Thus, CEC and Link 16 are discussed in more detail below. National information feeds (coming from the left-hand box) are also important, especially for cueing sensors. While there is no further detail to be presented here, it should be noted that the timeliness of delivering these data could stand improvement.

The quality of the situational picture derived from the input data is of course critical and is a matter of much concern. A new effort, the SIAP System Engineering Office Program, is being established to address this concern. In addition, the area air defense commander (AADC) module has been established to provide a display capability to help time-critical (but not real-time) decision making and non-real-time planning. Both the SIAP program and the AADC module are discussed in more detail below.

### 3.3.4 Link 16

Describing Link 16 is complicated because it is many things rolled in one—it describes an RF communications network architecture, provides a message set for conveying information through the network communications, and defines procedures for the way in which this information will be gathered.

#### 3.3.4.1 Current Capabilities

Link 16 describes a networking scheme and message set that are instantiated in radio terminals. The JTIDS and its slightly more modern variant, the multifunction information distribution system (MIDS), are the Navy's terminals of choice. These terminals will be installed on a variety of aircraft, surface ships, and submarines over the next several years, as well as in Patriot and THAAD forces. Original JTIDS development (and the corresponding Link 16 specification) dates back at least 30 years; thus, even though it is just being deployed now, it is very much a legacy capability.

Link 16 uses a time division multiple access (TDMA) networking scheme.[5] In the basic configuration, this means each participant on the net can transmit only in its allocated time slot and must be in receiving mode the rest of the time. If only one time slot is allocated to a participant, Link 16 will transmit once every 12 seconds. It is possible to establish multiple independent networks simultaneously by giving each net a different frequency hopping pattern for its transmissions. In general, the TDMA scheme is very complex to arrange and quite demanding on operator skills. Up to a week or two can be required to develop and test the scheme to be used in an actual operation. Thus, Link 16 does not currently support flexible, rapidly conceived operations.

The maximum capacity of the JTIDS (or MIDS) radio in antijam mode is 115 kbps (and often much less in practice),[6] a low figure by modern information transfer standards and one that limits the utility of JTIDS. This is significant, because DOD has mandated that JTIDS (or variants such as MIDS) will provide

[5]Details are given in Appendix C.

[6]Appendix B provides an analysis of capabilities and limitations of Link 16.

the basic tactical communications capability. A more modern approach based on commercial technology would appear to greatly increase bandwidth. As discussed in Appendix C, commercial wireless technology is advancing rapidly, and capacities of at least a few megabytes per second currently appear possible. The commercial technology appears to have the necessary quality of service for military applications, although jam resistance is not a significant factor in the commercial developments. Still, the commercial technology would offer a good base upon which to build a jam-resistant capability.

The Link 16 fixed-format message set—called the J-series messages—covers a wide range of information categories. Very important among these, of course, is the surveillance tracks detected by participants in a Link 16 net. To obtain the best data on a given target and avoid redundant tracks, Link 16 procedures call for the platform with the "best" track to have reporting responsibility and to be the only platform to report that track. In practice, this can lead to significant difficulties. Other message sets allow for mission assignment to attack a target, and still others provide precision position location information (PPLI) based on packet time-of-arrival measurements. This PPLI information allows for relative navigation and also serves as an identification means.

Since the Link 16 message set was developed in the context of air defense, it covers the sort of information needed for cruise missile defense. Ballistic missile defense, however, required new messages to be added—for example, messages referring to missile launch and predicted impact points, space tracks, and engagement status. These additional messages take a shoot-and-shout approach to ballistic missile defense, but they do not provide coordination among multiple platforms that could fire at a given ballistic missile.

### 3.3.4.2 Planned Improvements

Operational experience such as was gained in the Kosovo air war indicated significant shortcomings in TADIL operation. At times, significant portions of the air picture were missing because different tactical data links (Link 16 and others) would not interoperate with one another. Better TADIL network management is necessary. To promote that, the position of JICO has been established, and procedures for its operation have been defined.[7] In addition, there are plans to develop an automated tool to help the JICO in conducting network management. While these procedures and the automated tool, coupled with training for the individuals involved, should aid TADIL network management, they underscore the complexity of TADIL operation and the need to adopt more modern network technology allowing simpler management.

---

[7]Joint Staff. 2000. *Joint Data Network (JDN) Operations*, CJCSM 3115.01, The Pentagon, Washington D.C., September 1.

TABLE 3.2 Planned Improvements and Potential Benefits

| Planned Improvement | Potential Benefit |
|---|---|
| Dynamic network management system (DNMS) for Link 16 | Incremental increases in the flexibility of Link 16 networks, perhaps coupled with greater ease of planning and configuring such networks |
| Enhanced throughput | Higher bandwidth communications across Link 16 radio channels |
| Optimized relative navigation | More accurate relative position and time information for Link 16 platforms |
| Joint range extension, S-TADIL J | Increased ability to transmit J-series messages across non-JTIDS radio channels |
| Link 16/JVMF advanced concept technology demonstration | Gateways between Link 16 radios and their messages, on the one hand, and the Army's messaging system on the other |
| Link 16 missile and tactical terminal (LMT2)/TacLink weapons | Tactical command and position/location links to guided munitions |

SOURCE: Information derived from McCloud, Kenneth L., "PMW 159 Advanced Tactical Data Link Systems (ATDLS) Program Office," briefing to the committee on July 26, 2000, Space and Naval Warfare Systems Command (PMW 159A), Arlington, Va.

The advanced tactical data link systems (ATDLS) program office (SPAWAR PMW 159) develops improvements to Link 16 and related TADILs. These improvements are summarized in Table 3.2. In general, the committee supports these improvements, although it expresses some particular reservations in the more detailed discussion in Appendix B. These improvement programs have technical merit and are likely to provide substantial benefits to the Navy. However, they are best viewed as late-life upgrades to a system that is nearing the end of its technical life cycle.

Serious consideration needs to be given to a much more modern approach to tactical data links. Such an approach would use a well-defined layered structure, as in Internet technology, instead of mixing the distinct problems of radio frequency (RF) channel architecture and message format, as Link 16 has done. Such an approach would also build on the rapid advances now occurring in commercial wireless technology.

### 3.3.5 Single Integrated Air Picture

A SIAP is said to be the "product of fused, near-real-time and real-time data from multiple sensors to allow development of common, continuous, and unambiguous tracks of all airborne objects in the surveillance area."[8] This is the desired result of the integration of situation inputs in Figure 3.5, above. Such a result does not now pertain. Instead one finds missing tracks, multiple track designations for one object, track number swaps between objects, and object misidentification. These shortcomings have been manifest in real-world operations and detailed exercises such as the all -Service combat identification evaluation test (ASCIET) series.

The preceding section highlighted the problems of Link 16 with regard to network flexibility and capacity. These problems are partly the result of not obtaining a SIAP, but the set of causes is much larger and includes basic technical shortcomings, the inconsistent implementation of a technical capability across different platforms, and the absence of necessary procedures. The root causes of the problem are numerous and include the following:[9]

- Lack of a common time standard across the force,
- Inadequate and inconsistent navigation capability,
- Poor tracking performance and inaccurate assignment of track quality,
- Connectivity shortfalls,
- Failure to achieve a common geodetic coordinate frame,
- Differences in correlation/decorrelation algorithms,
- Differences in automated identification processing,
- Limited and inconsistent implementation of message standards,
- Shortfalls in joint tactics, techniques, and procedures, and
- Difficulties in network design and management.

To confront the problem, the JROC directed in March 2000 that a SIAP system engineering office be formed.[10] The SIAP system engineer is responsible for the systems engineering necessary to develop recommendations for systems and system components that collectively provide the ability to build and maintain a SIAP capability. By JROC direction, the Navy will provide the lead system engineer, the Air Force will provide the deputy lead engineer, and the Army will serve as acquisition executive.

The SIAP system engineer has emphasized the importance of establishing

---

[8]Joint Theater and Air and Missile Defense/Combat Identification Division (J85). 2000. *Theater Air and Missile Defense (TAMD) Capstone Requirements Document (CRD) (U),* Draft, U.S. Joint Forces Command, Norfolk, Va., June 15 (Classified).

[9]Wilson, CAPT Jeffery W., USN, "Single Integrated Air Picture (SIAP) System Engineering," briefing to the committee on August 30, 2000.

[10]While the committee was not briefed on the program, it should be noted that the Family of Interoperable Operational Pictures (FIOP) program being developed in the Office of the Undersecretary of Defense (Acquisition, Technology and Logistics) is addressing issues related to the SIAP effort.

the necessary system engineering process and not just isolated improvements. Thus far, the office has identified candidate solutions to address the root causes noted above. Near-term emphasis will be placed on engineering and recommending SIAP-related improvements to fielded systems—in particular, identifying fixes to the joint data network (JDN). The JDN is the network formed from tactical data links, which in the future will be dominated by Link 16 for U.S. forces (but will also contain Link 22 for NATO forces).[11]

The SIAP System Engineering Office was created to meet a critical need, and its activities thus far appear well directed. The committee believes that the Navy should support the activities of this office and monitor them to make sure they are meeting naval needs. The committee further believes that the SIAP system engineer should take an aggressive stance in promoting the development of modern alternatives that would eventually replace the current tactical data links.

## 3.3.6 Cooperative Engagement Capability

### 3.3.6.1 Planned Capability

CEC combines measurement-level data from multiple radars and other available sensors in near real time to form a composite track picture. The Navy's intent is to deploy CEC widely—on cruisers, carriers, some destroyers, amphibious ships, and surveillance aircraft.[12] Initial focus is on air defense (primarily ship self-defense against cruise missiles), but later developments will address ballistic missile defense.

The composite track picture provides each CEC participant with a better track picture than that participant could generate alone. For example, if a target is dropped by one radar, other radars can fill in, and target location can be determined more accurately by combining observations from multiple sensors. Furthermore, each participant has a larger battlespace picture, one that is produced by the combined coverage of all the sensors. This larger coverage will allow a given participant to launch its defensive missiles before its radar acquires a target—the so-called engage-on-remote and forward-pass concepts.

The heart of the CEC system is the cooperative engagement processor (CEP) and the data distribution system (DDS). The CEP located on each platform correlates all the sensor input to that platform to form the composite picture. The DDS effects the high-bandwidth radar data distribution among the partici-

[11]See CJCSM 3115.01 for more discussion of the JDN. Joint Staff. 2000. *Joint Data Network (JDN) Operations,* CJCSM 3115.01, The Pentagon, Washington, D.C., September 1.

[12]There are some funding difficulties, however. The Navy's POM-02 budget submission dropped funding for installing CEC on existing E-2C aircraft and included it only for new E-2Cs. Existing E-2Cs comprise the bulk of the planned E-2C force.

pants. A prioritization scheme has been developed to send the most relevant data to each participant within the limits of the available bandwidth.

Control of defensive missiles being fired using CEC data lies outside of CEC (in the C&D module in Figure 3.9). Thus, CEC does not form a complete BMC3 system, nor was it intended to; loosely speaking, it is a distributed sensor system.

Over the last half-dozen years or so the CEC components have been upgraded and modernized, taking advantage of advances in computer and electronics technology and making increased use of commercial components. The production of CEC components began in 1998 at a low rate. Currently, CEC version 2.1 is undergoing large-scale, at-sea testing (the so-called Underway series).[13] Operational evaluation is planned for the spring of 2001. Version 2.1 will provide an air-defense capability; ballistic missile defense capability is planned for version 2.2.

### 3.3.6.2 Possible Extensions

In a CEC system, large amounts of data are transferred on a point-to-point basis between nodes, so scalability is an issue—that is, whether adequate amounts of data can be transferred as additional nodes are added to the system. This is one of the issues that will be addressed, at least for modest-size configurations, in the Underway tests. Furthermore, a new concept, the tactical component network (TCN), has been proposed that claims much more efficient data transfer. If this capability were realized, it could mitigate any scalability problems or even—possibly—allow for reduced bandwidth connections. The Navy is planning to investigate TCN and will outfit two cruisers with the capability. At this time, however, the eventual utility of TCN cannot be reliably predicted.

The original concept for CEC was to enhance ship self-defense in carrier battle groups. Additional uses are being considered and warrant review here. The principal question is whether the extensive CEC capabilities are needed for these additional uses or whether lesser (and presumably less expensive) capabilities would suffice. CEC is being planned for use in naval ballistic missile defense and is also being considered for joint theater ballistic missile defense. However, a ballistic missile track picture is much easier to obtain than a low-altitude cruise missile picture.[14] The question is thus whether the exchange of

[13]For example, Underway 10, conducted in September 2000, involved six CEC-equipped ships (1 CVN, 4 CGs, and 1 LHD), two CEC-equipped aircraft (an E-2C and P-3), and three CEC land sites. BQM-34 drones were used as surrogates for cruise missile targets.

[14]In cruise missile defense, the target can maneuver and present rapidly changing RCS to radar. This results in dropouts of target tracks and stresses track initiation algorithms. Ballistic missiles generally follow a Newtonian trajectory.

track data over an improved tactical data link capability, as could possibly be realized through the SIAP program, would be adequate.

CEC is also being proposed for uses where there would be little overlap in coverage between sensors. For example, an advanced concept technology demonstration is exploring coupling Aegis and Patriot via CEC for low-altitude cruise missile defense. The main benefit of CEC appears to be that it provides a composite track picture from overlapping sensor coverage, in which instance it is valuable for exchanging measurement-level data. When the coverage regions do not overlap significantly, it could suffice just to send track data, which could be done via a (possibly enhanced) tactical data link.[15]

The committee believes CEC can provide a valuable capability for ship self-defense and overland cruise missile defense if adequate overland sensors are available in the latter case. The committee does not have adequate information to take a position on the issues of extended use noted in the last two paragraphs. However, it believes that the Department of the Navy and the joint community should conduct adequate analyses to resolve these issues if they have not already done so. No such analyses were apparent in the briefings received by the committee.

Rather, it appeared that since CEC was an existing capability, at least in prototype form, it was being extended to new uses without an adequate analysis of the alternatives and trade-offs involved. The advantages of using an enhanced tactical data link capability could be reduced cost and greater operational flexibility in passing the data, since tactical data link terminals will be more widely deployed. Furthermore, just as one should guard against locking into legacy Link 16 technology, one should also be cautious about locking into CEC technology. While CEC is highly capable, it must be kept in mind that it is based on an architecture first designed in the 1980s.

### 3.3.7 Area Air Defense Commander Module

Joint doctrine calls for the establishment of an AADC to oversee air defense operations under a joint task force commander.[16] The AADC module is a display capability and associated tools for use at the AADC (i.e., operational) level as well as at the tactical level. While Navy doctrine does not have an exact analogue of the AADC, the AADC module is intended for use in naval as well as joint operations.

[15]There could be an advantage to netting multiple Patriot systems together using CEC if there was significant coverage overlap among the Patriot radars.

[16]Ross, Lt Gen Walter K., USAF, Director, Joint Staff. 1996. *Doctrine for Joint Theater Missile Defense,* Joint Publication 3-01.5, The Pentagon, Washington, D.C., Available online at <http://www/dtic.mil/doctrine/jel/new_pubs/jp3_01_5.pdf>.

The AADC module display shows air and ballistic missile defense assets, hostile forces, and neutral entities—all depicted as the real objects in a three-dimensional representation instead of in terms of some abstract symbology. This information is updated through information feeds such as Link 16.

In joint operations, the joint force air component commander (JFACC) airspace control authority (ACA) prepares the airspace control order (ACO) that determines the partitioning of the airspace to deconflict the various offensive and defensive assets that will be operating in it. The AADC module provides a three-dimensional rendering of this partitioning. In addition, it displays such operational parameters as the coverage areas of surveillance systems and the range of weapon systems. It also supports collaborative planning by providing a visual teleconferencing capability.

The AADC supports both planning and execution. Its displays and tools allow the initial positioning of air defense forces to be determined much more rapidly than with the conventional manual procedures. However, the material presented to the committee on the AADC module did not appear to indicate that the operational concept for the interaction between the AADC and the JFACC had been fully worked out—for example, the concept for the coordination and airspace deconfliction of offensive and defensive operations, which is necessary to take full advantage of the AADC module's capability. Similarly, further development of the operational concepts for joint ballistic missile defense also appears to be required.

In execution, the AADC module's display and tools allow for the near-real-time tasking and redirection of defensive assets. This capability should aid tactical command and control of defensive operations significantly. The committee did not, however, receive adequate information to be able to assess the sufficiency of the AADC module's battle management tools. That is, while there is significant capability in the module now, further automated battle management aids could be desirable to cope with complex, multitarget situations.

There is an important cautionary note pertaining to accuracy: The AADC module's display is very realistic. Such displays can lead observers to believe that is how the real situation is, when in fact there can be errors in location, identification, and completeness in the data input to the display. Operators should guard against taking the displays more literally than is warranted. Means should be sought for depicting the uncertainties in the AADC displays. Furthermore, safeguards against the engagement of neutral targets—such as the inadvertent shooting down of an Iranian Airbus in the Gulf many years ago—must be incorporated in the system. For example, as currently envisaged, the AADC makes no use of the *Official Airline Guide,* and it has no links to civilian air traffic control.

AADC prototype modules have been installed on the command ship USS *Mount Whitney* and the cruiser USS *Shiloh.* The prototype module on the *Shiloh* was used in the rim of the Pacific (RIMPAC) exercises in the summer of 2000. Its use was apparently well received. Further testing of the AADC module is planned, for example in the *Abraham Lincoln* battle group, and initial operational capability (IOC) is planned for FY06.

In summary, the AADC module should provide a valuable capability supporting those management/control functions shown on the right-hand side of Figure 3.5, above, as well as the non-real-time planning function. However, as noted, further development of the operational concepts necessary to execute these functions could be warranted, and serious consideration needs to be given to the representation of uncertainty in the battlespace display.

### 3.3.8 BMC3 Summary

The BMC3 discussion above, augmented by the material in Appendix C, may be summarized in terms of a set of conclusions. Overall, the committee found that BMC3 concepts and technical capabilities require significant rethinking and development to meet missile defense needs. More specifically, the committee concluded as follows:

- *Operational concepts and the associated technical capabilities must be able to support highly adaptable missile defense force configurations; the current approach—thinking of prescribed configurations—is not adequate.* Experience has shown that force components must be pulled together in unplanned ways and unanticipated assets often added in. What is required is a technical basis that makes this jury-rigging readily accomplishable—namely, a network-centric architecture that allows the easy interconnection of assets and enables users to readily identify information and get it from any source. Current missile defense BMC3 architectures are not of this type.
- *Wide-area missile defense puts an increased premium on BMC3, to which current Department of the Navy efforts are not paying adequate attention. Furthermore, as the threat becomes more stressing, even local defense will require more emphasis on BMC3 to increase its horizon against threats.* For example, in wide-area mission overland cruise missile defense, naval forces lack effective surveillance capability and would need the capability provided by a platform such as the AWACS or, perhaps, a group of UAVs. Crossing Service lines like this means that the appropriate technical and procedural capabilities must be in place; the committee saw no evidence that these requirements were being addressed for overland cruise missile defense. Ship self-defense is an example of local defense where the threat is expected to increase in terms of both numbers and reduced detectability. Interfacing with an AWACS, for example, would increase the horizon, allowing the defense more time to meet the threat. The general point to be drawn from this is that effective future theater missile defense could require not only the physical distribution of sensing, control, and shooting assets, but also their distribution across Services. This would entail a major cultural change for traditional Service operations.

• *While both Link 16 and CEC provide valuable capabilities, neither is likely to be evolved far enough to provide the capability needed for flexible connectivity.* Link 16 does not allow for the rapid incorporation of unplanned platforms or unanticipated information sources into its network. Enhancements are planned, and they will be useful for near-term improvements to the network, but there are limits to these improvements given the very dated technology and architecture upon which Link 16 is built. CEC was designed to be a distributed radar, and it is quite effective in that sense. However, while it does have a high bandwidth, it was not designed to be a multipurpose communication system easily accommodating the inclusion of nodes not designed to its specialized interface specifications.

• *Newly emerging commercial wireless technology can be leveraged to meet missile defense communications needs.* Commercial technology is providing multi-megabit-per-second wireless communications and has developed quality-of-service capabilities and some information assurance capabilities. Although antijam capability is typically not a feature of commercial technology, that technology should nonetheless be a good starting point for adding in this capability. Current improvement efforts face the coupled problems of limited bandwidth and poor battlespace control capability. Solving the bandwidth problem disentangles the two problems and allows focusing on battlespace control.

• *Determining an accurate battlespace picture and coordinating the assets in it remains a difficult problem that requires much more attention.* Current efforts to improve battlespace coordination must be continued and augmented with more advanced research. Increased bandwidth will allow greater data exchange, which should allow better correlation of detections, but significant improvements beyond that will still be required. Programs such as the SIAP and FIOP are necessary, and even more advanced research programs are necessary. Areas of research include the decentralized management of resources and the management and presentation of uncertainty.

# 4

# Assessment of Current and Projected Department of the Navy, Other Service, and Defense Agency R&D Programs

## 4.1 NAVAL MISSILE DEFENSE R&D PROGRAMS

Research and development efforts related to ballistic missile defense fall under the budgetary jurisdiction of the BMDO. Consequently, the main thrusts of the Department of the Navy R&D programs are related to CMD. However, the Department of the Navy, primarily through ONR's missile defense[1]—and, to a lesser degree, its platform protection[2] future naval capability (FNC) efforts—is pursuing efforts that are relevant to both BMD and CMD in areas such as the following:

- IR sensors,
- Combat identification,
- Advanced ground-based radar technologies, including the advanced multifunction radio frequency system (AMRFS),
- Various critical radar components—for example, GaN and SiC microwave power amplifiers, and
- High-speed digital circuits.

---

[1]Cetel, CAPT Alan J., III, USN, "Missile Defense (MD) Future Naval Capability (FNC) Program Overview," briefing to the committee on April 26, 2000, Office of Naval Research (Code 35)/Office of the Chief of Naval Operations (Code 091), Washington, D.C.

[2]Lawrence, Joseph P., III, "Department of Navy S&T Platform Protection FNC," briefing to the committee on April 26, 2000, Naval Research Laboratory, Washington, D.C.

As currently prorated for the FY02 budget, the Department of the Navy's two FNC programs that are directed toward missile defense have 6.2 and 6.3 funding levels that total $70 million to $80 million per year. The objective of FNC programs is to focus Department of the Navy 6.2 and 6.3 funding to obtain a better return on investment in terms of fielded capabilities. Thus, projects are funded based on requirements, capability gaps, technology feasibility, transition availability, and program manager commitment. There is much to be said for this approach, but a shortcoming is that it tends to focus resources on evolutionary as opposed to revolutionary approaches. The latter are likely to be viewed as technologically risky, and it is intrinsically difficult to identify concrete transition paths for such approaches.

### 4.1.1 Department of the Navy Cruise Missile Defense Sensor Research and Development

As discussed in Section 3.1, which gives an overview of theater missile defense capability, the Navy is in the process of procuring and developing four major radar systems that have the potential of providing improved sensors for Navy ships. Accordingly, sensor R&D carried out under ONR's missile defense FNC is largely oriented to providing improved sensors for OCMD.

In tactical situations, overland cruise missiles are difficult to detect and track because of clutter from the land background. In many situations OCMD is complicated by the fact that the flight path of cruise missiles may be programmed to exploit terrain masking. Another complication is that the engagement may take place at ranges that are below the horizon of sea-based (and even some land-based) radars. Generally, a single land- or sea-based sensor will not allow robust acquisition of remote land-attack cruise missiles. The combined effects of terrain masking and radar horizon limitations necessitate one or more airborne AMTI radar platforms (e.g., JLENS, E-2C RMP, AWACS, JSTARS, and UAVs) or some other form of distributed cooperating short-range electro-optical, acoustic, RF, or other sensors.[3]

Appropriately, ONR's missile defense FNC program is concentrating on the elevated sensor problem. A sensor by itself does not constitute an OCMD system. Detections and tracks developed by an elevated sensor must be passed to a weapon release authority. If, based on detections by an elevated sensor, a weapon has been released, it must be guided into a collision course with the incoming missile. When the interceptor comes close enough to the target to allow its onboard sensor to acquire the target, the terminal encounter will occur autonomously. ONR's missile defense FNC is engaging in R&D efforts related to all phases of this problem.

---

[3]In Section 4.1.3.2, the Link 16 and CEC legacy discussion applies also to the problem of naval connectivity to these joint sensors.

Although clutter cancellation is the main issue in the development of AMTI radars, radiated system power and sensitivity are not irrelevant. Cruise missiles can and do have very low RCS values, particularly when viewed nose-on. Detection of low-RCS targets requires great system sensitivity. Unfortunately, the lower the RCS of the target to be detected, the lower the detection threshold must be set. Very low threshold levels result in the detection of many spurious targets (e.g., noise spikes, birds, and bugs). Robust discrimination algorithms must be developed that will reject these spurious detections efficiently and thereby minimize the computer resources needed to reject false targets. The development of false target rejection algorithms in order to permit operations at the low thresholds needed to counter low-RCS cruise missiles is certainly an appropriate area of activity for ONR's missile defense FNC.

Other possibilities for the detection of very low RCS objects include the use of multistatic radar configurations. Stealth technology generally reduces the amount of energy that is reflected back to a conventional monostatic radar. Energy reflected from a low-RCS target in other directions can be high, allowing the detection of strong glints by a properly positioned receiver that is not colocated with the radar transmitter. Multiple geometrically dispersed receivers must be available to increase the probability that at least one receiver will detect a strong glint, in effect increasing the target's RCS. The difficulties associated with multistatic operation are formidable. Some of these difficulties may be overcome by the application of current technology. Others will require an extensive R&D program. The committee is optimistic that the heretofore limiting difficulties associated with multistatic operation can be conquered and believes that R&D efforts in this area would be an appropriate component of ONR's missile defense FNC effort.

Another interesting possibility that might be included in ONR's missile defense FNC effort would be the exploitation, by means of image-processing techniques, of the target's obscuration of the background as revealed through its motion—that is, by imaging the target's moving RF shadow. This obscuration is determined by the physical extent of the object, not its apparent RCS.

After having sorted out the cruise missile from low-threshold-induced competing "targets," the cruise missile defense system must be capable of robust combat identification as part of the discrimination process, for the targeted object could be a friendly cruise missile or aircraft.

### 4.1.2 Department of the Navy Cruise Missile Defense Weapon R&D

Although a full-scale acquisition program apparently does not exist, it is clear that an OCMD concept based on weapon launch from a remote, sea-based platform will require a weapon with unique capabilities not represented in the Navy's current inventory of weapons. ONR managers are aware of this deficiency and are supporting the development of technologies that will enable the building of such weapons.

For cruise missile interceptors, target handover prior to terminal engagement is easier than target handover in ballistic missile defense. Typically, the threat is a single object not supported by external penetration aids. Under this condition, semiactive handover and terminal guidance, coupled with an active RF fuze, are adequate. This is the approach used by the Navy self- and area-defense systems today. In the OCMD situation, terrain masking and the effects Earth's curvature preclude the use of surface-based semiactive guidance in most cases. The committee believes that the ONR missile defense FNC should focus on the development of new techniques that will make surface-based semiactive guidance unnecessary.

The development of weapons to support ASCMD was discussed in Chapter 3. R&D for extending the capabilities of ASMD weapons is not a component of ONR's missile defense FNC. However, under ONR's reactive warhead program, a reactive fragmentation warhead is being developed for transition to a number of possible ASCMD and OCMD interceptors.

Under associated programs, R&D for the development of improved electronic warfare techniques is being pursued. Based on briefings provided to it, the committee perceives that the Navy is continuing its impressive program of finding novel extensions for traditional EW techniques. This effort appears to be well funded and is apparently resulting in the near-term deployment of new and highly effective EW capabilities.

At the time of this study, the Navy did not have a program of record for laser or directed-energy weapons. Although it worked intensively for about 30 years on the development of such weapons, no system achieved operational status.

As discussed in Section 4.3, on Air Force R&D programs for missile defense, advances in the technology for the chemical oxygen-iodine laser (COIL) and free electron laser (FEL) show some promise. However, given the present status of these technologies and the lack of a Navy program of record to support them, the committee believes it is unlikely that any laser or directed-energy weapons will achieve IOC on Navy platforms before 2015 or 2020.

### 4.1.3 Department of the Navy BMC3 Research and Development

#### 4.1.3.1 Background

BMC3 provides the glue for connecting weapons to sensors in missile defense systems. Research and development in BMC3 algorithms, software, processors, and communications is required to outpace the threat, to develop lower-cost solutions, and to provide the flexibility to tie together evolving sensors and weapons—naval and other U.S. Services and allied—into a coherent system of systems.

In consideration of the technology needs for missile defense BMC3, the committee finds it useful to distinguish between the BMC3 algorithms per se and the communications links and networks, processors, and software required to implement these algorithms as a distributed information processing system. The requirements for the former are relatively unique to missile defense, while the requirements for the latter strongly overlap those for commercial information processing systems.

Achieving a high-probability kill of a TBM is a difficult problem, so lethality is a key concern for BMD. BMD systems have traditionally been structured in many layers to achieve a cumulative probability of kill exceeding that of any individual layer. The BMDO TMD family of systems has been structured in this way, with the THAAD system and the NTW system providing overlays for the PAC-3 and the NAD system. The Air Force's airborne laser (ABL) could provide a boost-phase layer for shorter range threats.

Exploiting the capabilities of multiple defensive layers in an expeditionary environment requires algorithms for coordinated, distributed weapon-target assignment. Although there are significant CONOPS issues, the development of an appropriate technology base could clarify the trade-offs between coordinated and completely decentralized engagement strategies.

For BMD, discrimination is a key concern,[4] and much effort has been devoted to the development of discrimination algorithms for both radar and optical sensors. These algorithms typically extract features from single-sensor data and partition feature space into regions characteristic of reentry vehicles, decoys, and other objects. Extensive training data are needed to select appropriate features and to define these partitions.

The major TBMD systems currently under development have both radar and optical sensors, and the potential benefits of combining the data they collect on various features of the threat objects is becoming increasingly evident. X-band radars being developed allow for the precision measurement of microdynamic features of threat objects.[5] The passive IR sensors being developed for performing onboard interceptor functions are naturally adept at measuring the thermal characteristics of threat objects. In addition, there is a large class of features, such as macrodynamic body motions, that both sensors can measure. The potential for significant improvements in discrimination capability lies in the effective fusion of these feature vectors.

---

[4]Although conceptually a key element of the BMC3 system, discrimination is often associated with sensor and/or interceptor technology. The committee discusses discrimination in this section, recognizing that discrimination algorithms may be physically hosted on a sensor platform or an interceptor.

[5]Microdynamic features of threat objects refer to spin rates or any other irregular motions that provide a unique signature that allows a RV to be discriminated from decoys and debris.

The limited amount of research in progress on fusing electro-optical and radio frequency sensors can use X-band radar measurements of a target's wobble or nutation in combination with optical measurements of radiant intensity to discriminate the target from replica decoys. The combination of passive optics and LIDAR is also being looked at for similar dual-sensor discrimination modes. There is also extensive investigation of using dual-phenomenology observations to mitigate the effects of various countermeasures; an example is the use of optical sensors to compensate for radar degradation caused by jammers and chaff.

One area where dual phenomenology and electro-optical/RF fusion cannot be implemented in the near term—even though they are clearly needed—is precommitment discrimination (before launch of an interceptor). Precommitment discrimination is needed to allocate and designate interceptors efficiently, but it will not be available until the space-based infrared system-low (SBIRS-low) is deployed.

Unfortunately, with the proliferation of ballistic missile technology, the likelihood of collecting the data needed to train the current generation of BMD discrimination algorithms is diminishing significantly. What is needed is a new generation of discrimination algorithms that reason based on an understanding of sensor and ballistic missile phenomenology. Unlike the current generation of algorithms, such algorithms would be able to cope with new objects and deployment mechanisms for which they have not been explicitly trained. Humans (e.g., missile test analysts) are able to operate in this fashion, but a huge research effort would be required to develop the algorithm technology that would allow discrimination to be automated.

Tracking and identifying cruise missiles in an overland environment with clutter and terrain masking has always been a difficult problem. With the proliferation of signature reduction technology, cruise missiles can defeat current systems. Improved sensor technology is needed to provide a low-cost, distributed sensor network, including bistatic radars and other novel sensing means to obtain track and identification data. Tracking and classification/identification algorithms are needed to exploit data from the sensors. These algorithms must fuse data from multiple sensors, incorporating a knowledge of the terrain and hypothesized missile objectives to extrapolate through coverage gaps. Sensor resource management algorithms are needed to ensure the operation of the sensor network as an integrated sensing system. Sensors must be positioned and tasked to provide assured detection of new threats while supporting the engagement of already detected threats.

Assigning weapons to targets in such an environment has complexities beyond those of the already-difficult BMD problem. Since cruise missiles fly in the same altitude regime as aircraft, UAVs, and certain friendly weapons, real-time, dynamic airspace deconfliction is necessary. Owing to terrain obscuration, it may not be possible to assure the continuous, fire-control-quality track of cruise missiles everywhere, so that engagement areas will have to be selected where sensors can support interceptor requirements. With concepts such as the air-directed surface-to-air missile (ADSAM), the platform providing interceptor support may not be the same platform that is launching the interceptor. As a result, there is a need for distributed algorithms to optimally coordinate weapon, sensor, and communications resources to defeat a low-signature cruise missile threat.

Algorithms for missile defense BMC3 are implemented in software on processors tied together by communications links. The software, processing, and communications technologies can heavily leverage commercial technology.

Software issues for missile defense BMC3 include real-time, secure, large-scale, adaptive, distributed processing. While these are areas of intense commercial interest, there are no completely satisfactory solutions available, as evidenced by the difficulties being experienced in current software-intensive DOD programs despite their extensive use of commercial off-the-shelf (COTS) software. Technology is needed to permit the integration of heterogeneous, independently evolving software components. Rigid interface formats and database schemas must be avoided in favor of technologies that permit interface and schema extension and evolution without modifying components that do not use the new data elements that are added. Distributed control mechanisms are needed to quickly reconfigure and execute software components in response to changing environmental conditions.

Commercial processing technology is directly applicable to missile defense BMC3 processing requirements. Where necessary, radiation-hardened versions of commercial systems can be employed. Thus, the development of special-purpose data processors for missile defense BMC3 is unnecessary in general.

As is described in Appendix C, commercial wireless communications technology is an area of great technological ferment. Commercial technology is available in the form of wide-bandwidth radio links and quality-of-service-enabled Internet equipment that more than satisfies military requirements for latency, message loss, bandwidth, and information assurance. Thus, the communications requirements for missile defense BMC3 could be best met by adapting commercial wireless networking technology and equipment (for example, by increasing its jam resistance).

#### 4.1.3.2 Department of the Navy BMC3 Technology Programs

***Theater Ballistic Missile Defense***

Development of advanced technology for TMD and BMC3 is the responsibility of BMDO. BMDO-sponsored work in BMC3 is discussed briefly in Section 4.4.4, "BMDO and DARPA BMC3 Efforts."

### *Cruise Missile Defense*

The Department of the Navy's efforts in cruise missile defense come under its missile defense FNC and platform protection FNC programs. A significant portion of the missile defense FNC program is addressing BMC3 issues.

In the platform multisensor integration (MSI) program, a sensor fusion capability is being developed for the E-2C. This capability will correlate and fuse radar, infrared, and electronic support measure (ESM) data to better identify and track targets.

The objective of the composite combat identification (CCID) program is to attach high-confidence identification to theater-wide tracks to get theater-wide combat identification (CID). The approach is to develop a universal CID engine that will collect CID attributes from all relevant sources in theater, correlate CID attributes to a common track database, reason over the data collected to produce high-confidence CID, and deliver CID with low latency to theater units.

The theater collaborative tracking (TCT) program is developing technology for a theater-wide tracking network that would improve bandwidth efficiency and reduce life-cycle costs (by eliminating the need to modify computer software as new sensors are added to the network). The program objectives are to develop and demonstrate a collaborative tracking architecture and algorithms that incorporate need-based data distribution, that have minimal bandwidth increase when participants are added to the network that require no a priori knowledge of sensor or data source location, that require no software changes to accept new sensors, and that include sensor resource management algorithms.

In the threat evaluation and weapon assignment (TEWA) program, algorithms are being developed for force-level TEWA in a distributed environment. These algorithms would perform automated threat evaluations that consider all air and missile threats and all assets requiring protection in the theater and then provide automated shooter and weapon recommendations that consider all potential combinations.

Work under the platform protection FNC program is focused at the platform level, so there are no BMC3 projects planned. There is an unfunded demonstration program called the horizon extension platform. It would demonstrate a small, long-endurance, tethered hovering platform with electrical power as well as optical fibers provided by the tether. Were this concept to be developed, it could provide a platform for a communications relay for missile defense BMC3 systems and perhaps for look-down sensors as well.

### *BMC3 Technologies*

Although the committee did not perform an in-depth analysis of the research programs reviewed so briefly in the preceding section, its general impression is that these programs are addressing many of the key BMC3 technology priorities.

The importance of discrimination algorithms in TMD and distributed tracking and of resource allocation algorithms in CMD is reiterated.

A concern for both TMD and CMD is the lack of a systems context and an evaluation test bed for BMC3 technologies, such as were developed for national missile defense BMC3 in the early to mid-1990s. For example, the experiment version-88 (EV-88) prototype BMC3 system and associated test bed developed by the Army in Huntsville, Alabama, and the space-based experimental version (SBEV) developed by the Air Force Electronic Systems Center. These test beds provided a means of integrating technology developed by multiple contractors, evaluating the contemporaneous COTS software technologies, and demonstrating the technology to users in a system context. They provided the basis for and led directly into the ongoing development of the national missile defense (NMD) BMC3 system.

The committee believes that a missile defense BMC3 test bed should be established.[6] This test bed would allow multiple participant organizations to demonstrate their technologies in a system context. The system concept should be relatively unconstrained by current military implementation considerations and CONOPS. Thus commercial wireless communications links and Internet networking technology should be applied. Commercial software technologies should be used in exercises to demonstrate the rapid integration of heterogeneous applications softwares to create a real-time, distributed BMC3 system. The focus would be on future threats, weapons, and sensor systems. This test bed would permit advanced technology to be evaluated in a low-cost environment incorporating it in a development program.

### *Approaches for the Provision of Improved BMC3 Capabilities for Naval Forces*

The committee believes that BMC3 for TMD will need to undergo a revolutionary redesign. Traditional approaches to BMC3 adopt a design philosophy that is overly static given the highly dynamic environment that will characterize ballistic missile and cruise missile defense. Systems that use these approaches directly link specific preplanned sensors to interceptors, creating a closed-loop control system that guides the interceptor to the target. The fatal flaw of such systems is that a failure, weakness, or unavailability of key components may

[6]A previous Naval Studies Board report recommended that Internet Protocol ports would provide valuable evolutionary enhancements (e.g., increased interoperability) and should be pursued (see Naval Studies Board, National Research Council, 2000, *Network-Centric Naval Forces: A Transition Strategy for Enhancing Operational Capabilities,* National Academy Press, Washington, D.C.). This committee agrees with that approach; however, it believes that merely pursuing the "wrapping" of legacy applications will not get the Navy to the desired modern end state—hence the emphasis on a test bed.

irreversibly degrade the entire system, and redundancy is needed to overcome this flaw.

The alternative that the committee proposes is to engineer a highly dynamic system in which the entire sensor-to-shooter chain is assembled in real time from whatever components happen to be working and available, very much in the spirit of network-centric operations.[7] In network-centric operations, information would be shared across the sensing network. Specific sensors would be brought into play and focused on a specific task if they can provide discriminatory power, and airspace is managed dynamically to allow the best use of sensors and the clearest paths for interceptors. Finally, interceptors would be tasked dynamically to afford the most effective protection for the most valuable assets.

## 4.2 ARMY MISSILE DEFENSE R&D PROGRAMS

### 4.2.1 Theater High Altitude Area Defense System

The Army's THAAD system (Figure 4.1) is designed to provide broad area coverage and a deep battle space against short-, medium-, and long-range theater ballistic missiles. It has the unique capability to engage targets at both exo-atmospheric and endo-atmospheric altitudes, giving it enough battle space to achieve multiple shots (the shoot-look-shoot firing doctrine). The SLS capability provides low leakage and minimal expenditure of interceptors. THAAD is an HTK defense system intended to provide high lethality against the full range of theater missiles in the current and projected threat over a wide range of crossing angles. The HTK strategy results in a relatively lightweight interceptor capable of reaching high burnout velocities and providing high firepower per battery.

THAAD may be characterized as an upper-tier system in that it has the reach to engage targets at high altitudes in either an autonomous or a layered defense mode. In a layered defense mode, operating cooperatively with a lower-tier system such as PAC-3, it can provide the first filter of a flexible, low-leakage defense in depth. With its high-performance, X-band radar, THAAD can perform kill assessment for either a second upper-tier shot or handover to a lower tier.

The THAAD missile has a single-stage, solid-propellant booster with thrust vector control and a separating kill vehicle. The kill vehicle employs a gimbal-mounted infrared seeker, providing precision target imagery for tracking and aim-point selection; an uncooled sapphire window; and a liquid, bipropellant

[7]Network-centric operations are military operations that exploit state-of-the art information and networking technology to integrate widely dispersed human decision makers, situational and targeting sensors, and forces and weapons into a highly adaptive, comprehensive system to achieve unprecedented mission effectiveness. See Naval Studies Board, National Research Council. 2000. *Network-Centric Naval Forces: A Transition Strategy for Enhancing Operational Capabilities.* National Academy Press, Washington, D.C., p. 1.

- Endo-atmospheric and exo-atmospheric upper-tier system
- X-band phased-array radar
- Passive MWIR optical homing guidance
- 8 to 10 missiles/launcher, 9 launchers/battery
- C-1 FUE, 2007, C-2 MR, 2012
- 2 consecutive HTK successes on 6/10/99 and 8/2/99, EMD contract on 6/28/00

FIGURE 4.1 Theater high altitude area defense (THAAD) system.

DACS. Autonomous onboard navigation, guidance, and target tracking are performed with in-flight updates from the radar. The seeker has a cooled focal plane array with MWIR indium antimonide (InSb) detectors.

Unlike the PAC-3 system, the THAAD system does not have a long history of evolutionary development. It entered the demonstration/validation (Dem/Val) phase of development in 1992. With the experiences and lessons of Desert Storm lending urgency to the development of improved TMD systems, THAAD embarked on an aggressive development schedule. The Dem/Val phase included an objective of first flight in 2 years, and delivery of a user OPEVAL system was targeted for 4 years after first flight. The program experienced a number of quality control and reliability problems in the flight test program, resulting in six consecutive failures to achieve target kill. No two of the failures were for the same reason, and none of them were related to the high-technology features of the system.

Following a period of internal and red-team reviews, the flight test program was resumed and two successive HTK intercept tests were successful. These flights, on June 10, 1999, and August 2, 1999, demonstrated that the basic design is sound and the HTK strategy is technically feasible. After the two flight tests, THAAD was approved for milestone II and satisfied the exit criteria for the program definition and risk reduction phase. On June 28, 2000, the Army Space and Missile Defense Command awarded the THAAD engineering and manufacturing development (EMD) contract to Lockheed Martin.

The THAAD GBR, formerly an independent radar development project, has become an integral part of the THAAD program and provides surveillance and fire-control support to the system. A large power-aperture, X-band radar provides the long-range search, tracking, and discrimination capability to fully support the fly-out and intercept capability of the missile. It has a single face and does not search all azimuths. The radar incorporates a rich repertoire of waveforms and algorithms to provide the precision tracking and discrimination required to meet the kill probability and coverage objectives of the system. The radar is capable of microdynamic and imaging measurements of objects in the threat cloud to effect precision discrimination, and its broad bandwidth, in the gigahertz range, provides the range resolution to execute such functions as length measurement for discrimination. The radar also tracks the THAAD missile, providing in-flight target updates and a TOM to the kill vehicle. The primary means of performing target kill assessment, critical to SLS and handover decisions, is radar observations.

The engagement sequence for the THAAD system is as follows:

- Radar detection (either autonomously or using external cue),
- Radar track of the target complex,
  - — Discrimination,
  - — Missile commitment,
  - — Missile inertial guidance in midcourse flight,
  - — In-flight target updates from GBR to the missile,
- Onboard seeker acquisition,
- TOM from GBR to missile,
- Onboard target designation and aim-point selection,
- Endgame homing and intercept, and
- Radar kill assessment.

The THAAD system is being developed in two configurations. The first configuration, C-1, will be developed to meet the key performance parameters specified in THAAD's ORD. This phase of development will demonstrate design and operational capabilities through a series of ground and flight tests, qualify the system to enter production, and validate system-manufacturing processes through low-rate initial production. The C-1 configuration will provide a substantial war-fighting capability, with delivery of a complete THAAD first unit equipped (FUE) system to the Army by 2007. The second configuration, C-2, will incorporate software enhancements to (1) enable full ORD compliance, (2) keep pace with evolving threats, and (3) apply lessons learned from operational experience. The C-2 materiel release occurs in 2012. The C-1 FUE will include 16 missiles, 1 radar, 2 launchers, and 1 BMC3I subsystem. The C-1 early operational capability (EOC), scheduled for early FY09, will include 48 missiles, 1 radar, 6 launchers, and 1 BMC3 subsystem.

Milestone III for THAAD is scheduled for the start of FY09, and full-rate production (FRP) planning is for 1,250 missiles, 10 radars, 76 launchers, 38 tactical operations stations/launch control stations, and 68 system support groups.

Some of the advanced technology candidates being considered for upgrades to THAAD, along with their potential contributions to meeting advanced threats, are the following:

- *Two-color MWIR/LWIR focal plane array (FPA).* Extending the seeker FPA spectral band from MWIR only to both MWIR and LWIR will increase the sensitivity of the seeker and improve onboard discrimination capability.
- *Interferometric fiber optics gyro (IFOG) for the onboard inertial measurement unit (IMU).* Improved IMU performance, already base-lined for NTW, is configured for PAC-3 insertion in production for tactical missile systems such as AMRAAM. It will be upgradable to the microelectromechanical system (MEMS) during spiral development.
- *Electro-optics/RF fusion.* Fusing the electro-optics/RF measurements of threat objects will improve discrimination capability and reduce dependence on a priori threat data.
- *Advanced windows.* Improvement in the aerothermal characteristics of the window will allow operation at lower altitudes and thus extend the available battle space; it will also speed up the difficult manufacture of the window, making the process not easier but faster.
- *Gallium nitride power amplifier for the GBR.* Gallium nitride offers an 8:1 power density improvement and a 2:1 efficiency increase over the gallium arsenide material currently in use. This improvement can be applied to detect smaller signature targets at longer ranges, thereby increasing battle space or decreasing the size and weight of the radar for the same performance.

### 4.2.2 Patriot PAC-3

PAC-3 (Figure 4.2) is the latest of three upgrades to the Patriot air defense system to provide a robust capability against theater ballistic missiles. The PAC-1/PAC-2 system, used in the Gulf War against Scud ballistic missiles, demonstrated a threshold level of capability against this threat, but a need for improved capability was evident. The PAC-1 and PAC-2 modifications to

- Endo-atmospheric, lower-tier TMD system
- C-band phased-array radar
- RF active homing guidance
- 16 PAC-3 missiles/launcher
- FUE September 2001, IOC 2005
- 5 for 5 successful HTK intercepts, 3 vs. ballistic missile, 2 vs. cruise missile targets (as of 9/00)

FIGURE 4.2 PAC-3 system.

Patriot, largely in radar coverage and missile warhead design, were made only a short time before the Gulf War, leaving little time for test and evaluation before they were used in combat. While the ability of the system to intercept Scud missiles was demonstrated, including field upgrades performed to cope with tumbling missiles, it was not clear if there were any warhead kills, so further improvements were warranted.

The need for a theater ballistic missile defense capability has been recognized by the Army for several decades, and the quest for a defense system to meet this need has been marked by a number of shifts between single- and multiple-mission approaches. In the late 1950s and early 1960s, the field army ballistic missile defense system (FABMDS), a self-contained, mobile defense system designed expressly to engage free-rocket-over-ground (FROG)-type ballistic missiles (Soviet short-range ballistic missiles), was under development, but it was phased out largely because of the difficulties of packaging a complete system in a single vehicle. Also, FABMDs gave way to a shift in Army priorities from TMD to air defense, the mission objective of a study conducted from 1963 to 1965 called Air Defense Systems of the 1970s (AADS70).

The SAM-D system, which was the product of AADS70 studies, began with a requirement for a dual-mode capability (TMD and air defense). To reduce its cost, the SAM-D system was reoriented to a single mission, air defense, in the early 1970s, and its name was changed to Patriot. It remained a single-mission air-defense system until the PAC-1 and PAC-2 modifications were incorporated just prior to Desert Storm.

The ABM Treaty, signed in May 1972, prohibited the upgrade of such systems to provide an ABM mode. While the treaty does not proscribe TMD system development or deployment, the ambiguities and controversies surrounding the distinction between TMD and ABM systems inhibited the development of that class of system for a number of years.

The main elements of a PAC-3 battery are a radar set, an engagement control station, and a launch station. The launch station consists of a mobile launcher carrying 16 PAC-3 missiles. In the basic battery, launchers can be located up to 10 km from the engagement control station. With the remote launch communication enhancements upgrade, currently under development, launchers can be located up to 30 km from the basic battery, thus extending the TBM-defended area significantly. The radar is a mobile, multifunction, phased array operating at C-band.

In a modern Patriot battery, there are 8 launchers, 4 of which are loaded with 16 PAC-3 missiles each (total of 64 PAC-3s) and 4 of which are loaded with 4 PAC-2 missiles each (total of 16 PAC-2 missiles). The PAC-2 missiles, originally designed to enhance TBM lethality through the use of large fragment size, are now inventoried for use against all classes of targets. The mixed inventory of PAC-2 and PAC-3 missiles gives the Patriot battery flexibility in engagement of ballistic and air-supported targets.

The original guidance system for the Patriot air defense, still used in PAC-2, was RF semiactive homing, with a downlink to allow implementation of target-via-the-missile (TVM) processing. The TVM approach was initially selected largely because the onboard computers did not have the required throughput, a limitation that has been diminishing rapidly with the march of Moore's law. The PAC-3 missile uses a $K_a$-band active seeker for endgame homing. This guidance system was developed and demonstrated in the experimental extended-range interceptor (ERINT) program, culminating in three consecutive hit-to-kill intercepts, before transitioning to the Patriot system.

Extensive design trade-off analyses were conducted between a semiactive and an active RF guidance mode and the active seeker before final selection of the active mode for PAC-3. The main factor leading to selection of the active RF guidance mode was its demonstrated hit-to-kill lethality in an endo-atmospheric environment. Since the hit-to-kill strategy obviates a warhead, the missile is smaller and lighter, allowing a larger number of missiles per launcher (four times as many as PAC-2). This increase in firepower is a significant factor in handling large raid sizes and in implementing a salvo firing doctrine to improve kill probabilities.

The engagement sequence of PAC-3 is (1) inertial fly-out of the missile following initial detection and tracking by the radar to a nominal intercept point in space, (2) onboard seeker acquisition, (3) midcourse homing using rapid-response attitude control thrusters, and (4) endgame homing to achieve hit-to-kill of the target. Precommitment discrimination is performed by the radar, including a high-resolution waveform that enhances discrimination performance and provides growth options for non-TBM target classification. A critical onboard function is aim-point selection to assure warhead kill, a function that was not accurately executable by Patriot during the Gulf War. Aim-point selection has been effectively executed in the PAC-3 tests conducted to date.

As noted, PAC-3 is an endo-atmospheric, or lower-tier, TBM system, comparable in altitude operating regime to the NAD system. The Army has analyzed PAC-3 from both an effectiveness and an operational viewpoint as an underlay to THAAD in a layered configuration, as well as an autonomous TMD system. Since the elements of the PAC-3 system are separate and distinct from those of THAAD, they can provide a statistically independent tier of defense yielding very low overall leakage. With a lower tier having 20 percent leakage, net system leakage will be 4 percent ($0.2 \times 0.2 = 0.04$). For the defense of high-value theater targets, this layered defense mode can be of immense value in providing a level of protection unachievable with a single system.

The PAC-3 system has had five consecutive successful tests, three tests against ballistic targets and two tests against cruise missiles. The active RF mode has proved to be effective against both classes of targets. Significantly, the tests conducted thus far demonstrate warhead kill against unitary warheads and high lethality against multiple canister warheads.

The PAC-3 system is currently in a low rate initial production phase, with 16 missiles scheduled for the end of FY01 and 32 in FY02. The full system FUE is scheduled for the fourth quarter of FY01, coincident with FUE for the PAC-3 missile. The PAC-3 IOC is scheduled for 2006.

Candidate technologies for block upgrades to Patriot PAC-3 include the following:

- *Solid-state transmitter.* Building on prototype development for PAC-3 performed in the atmospheric interceptor technology (AIT) program, this program would transition a K-band solid-state transmitter to production. It would provide an alternative to the traveling-wave-tube-based transmitter. The program includes producibility enhancements, consolidation of the support electronics through the use of application-specific integrated circuits (ASICs), and environmental stress testing of a prototype.

- *IFOG.* A COTS IFOG unit has already been configured for PAC-3 insertion, and it would be a relatively straightforward upgrade to improve navigational accuracy with lightweight, low-cost gyroscopes. An opportunity for further improvement exists in the MEMS implementation of an IMU, a design not expected to reach maturity until at least 2005 or so.
- *Upgrades either under way or being evaluated for the Patriot radar.* These include advanced A/D converters, advanced digital signal processors, and improved discrimination and classification. A recent example of improved discrimination was the successful demonstration of a wideband frequency-jump burst waveform to measure body length.

### 4.2.3 Joint Land Attack Cruise Missile Defense Elevated and Netted Sensors

As shown in Figure 4.3, JLENS is a theater-based system employing advanced sensor and networking technologies to provide wide-area surveillance and precision tracking with a focus on land attack cruise missile defense. The role of JLENS is to expand the battlefield commander's surveillance and engagement capability against cruise missiles and other targets by extending the battle space for systems such as Patriot, MEADS, SM, and AMRAAM. JLENS can conduct surveillance between 250 and 300 km and fire control between 125 and 150 km from altitudes of up to 15,000 feet. JLENS can be based on land or sea and is tactically relocatable and transportable in-theater by C-5 or C-17 transport aircraft.

The JLENS primary sensors consist of two radars (one for surveillance and one for precision tracking and illumination) flying on two 71-m aerostats (unmanned, tethered, nonrigid aerodynamic structures filled with helium and air), each having a powered fiber-optic tether. Depending on the employment concept, JLENS uses a relocatable mooring system or a mobile ground-mooring station to launch, maintain, and recover the aerostat.

A ground-processing station controls the air vehicle, radar operation, and dissemination of location and tracking information to air and missile defense (AMD) BMC3 nodes and weapon systems. A processing station, configured in a transportable shelter, will be associated with each aerostat. Unique control interfaces coordinate operations between the precision track illumination radar and surveillance radar to execute and develop a SIAP.

JLENS provides over-the-horizon, 360-deg surveillance and precision fire control data for AMD systems. From its position above the battlefield, JLENS detects and tracks targets normally masked from a ground-based sensor. Tracking incoming cruise missiles allows their engagement by surface-based AMD systems, typically beyond the horizon, well before organic system radars can see

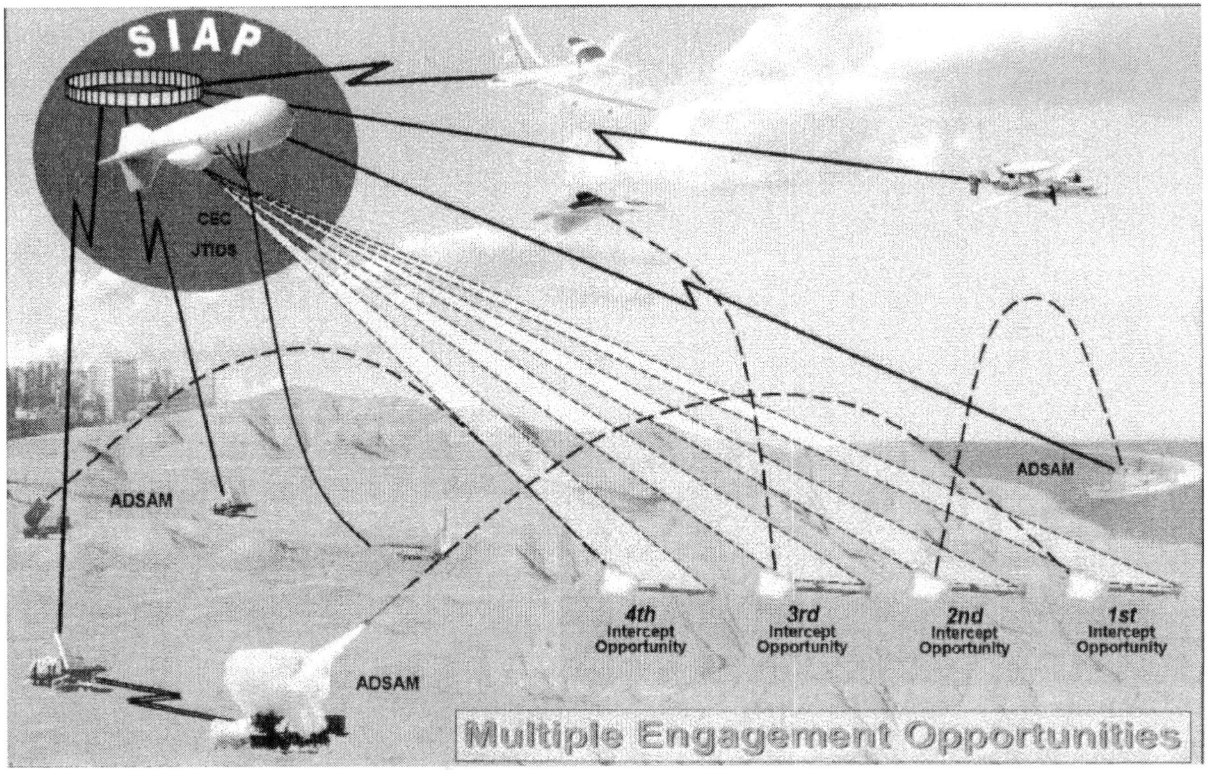


FIGURE 4.3 Joint land attack cruise missile defense elevated netted sensors (JLENS).

the targets. JLENS can stay aloft for up to 30 days, providing 24-hour coverage over extended areas, and it assists in the development of the SIAP by integrating data from multiple sensors.

JLENS distributes surveillance information via the joint data net and is netted with other theater sensors for distribution of fire control quality data. This netting will initially be accomplished via the CEC network and the joint composite tracking net. The CEC will fuse measurement data from JLENS sensors with data from other CEC-integrated land, sea, air, and space sensors to facilitate development of a SIAP and to provide early warning, cueing, and fire control quality data for over-the-horizon/non-line-of-sight engagements. The JLENS classification, discrimination, and identification data will also be fused via composite identification processing to support identification determinations in distributed command and control nodes.

In addition, JLENS is designed to support attack operation and communication missions. JLENS provides battlefield commanders with surface-moving target tracking and identification to support engagements by attack operation weapons. It also provides the basis for vectoring aircraft to intercept hostile aircraft while still over the horizon. The system assists in maintaining total situational awareness. JLENS has a demonstrated capability to elevate tactical communications and data networks above the battlefield to provide extended range and reliable connectivity and relay capabilities.

JLENS will use a blocked acquisition approach. Block I will design, fabricate, test, produce, and deploy the fire control radar, with sector surveillance integrated into the 71-m aerostat along with the processing station and ancillary equipment. Block II will similarly develop, procure, and deploy the surveillance radar. Block I and Block II configurations will meet the ORD threshold system requirements. Block III will consist of preplanned product improvements to develop and incorporate advanced technologies into Blocks I and II and to bring the JLENS system into compliance with the ORD.

In the FY02 to FY07 program objective memorandum, JLENS is in a program-definition and risk-reduction phase at the start of EMD. It is planned that 12 JLENS units will be built during this phase.

## 4.3 AIR FORCE MISSILE DEFENSE R&D PROGRAMS

The committee did not have an opportunity to review all R&D programs sponsored by the Air Force that might be relevant to theater missile defense. However, a number of committee members have had, in contexts unrelated to this study, extensive interactions with Air Force R&D programs, and they are aware that the Air Force is undertaking efforts in three areas:

- Improved sensors for TMD applications (e.g., improved performance of AMTI, GMTI, and electro-optical/IR sensors),

- Improved BMC3 systems for TMD applications (e.g., weapon control systems), and
- Improved weapons for TMD applications (e.g., improved performance of air-to-air missiles, EW techniques, and laser weapons).

Because not all of the committee members had interactions with the Air Force R&D programs, only the programs that were briefed to the entire committee by Air Force representatives will be covered here. These programs are related to laser weapon developments and are most significant in the context of the Navy's NTW TBMD effort.

The Air Force's BMD program is focused on the ABL project. The proposed ABL system will consist of a multimegawatt laser carried aboard a modified Boeing 747 aircraft. The system is designed to engage TBMs during their boost phase at standoff ranges of several hundred kilometers. In theater, the ABL will not require penetration into enemy airspace and will be able to engage the shorter range threats. Nevertheless, it will possess a self-defense suite. Furthermore, although its primary mission will be missile defense, the system, by its nature, also opens opportunities for applications in other missions. These might include the following:

- The protection of high-valued airborne assets against surface-to-air missiles,
- The accurate determination of launch points, and
- The collection of postboost tracking data that would provide cues to the other BMD systems to enhance their performance.

The ABL program requires integrating a multi-megawatt COIL into the aircraft to kill boosting TBMs. The ABL laser system consists of three main segments:

- A laser segment to provide laser power;
- A beam control/fire control segment to acquire the target, align the laser, compensate for atmospheric distortion, and propagate the laser beam through the nose-mounted turret; and
- A BMC3 segment to provide surveillance, communication, planning, and central command and control of the ABL weapon system.

The turret assembly contains a 1.5-m-diameter primary mirror mounted on the nose of the aircraft. Six onboard infrared sensors will provide 360 deg of coverage to permit autonomous detection of boosting missiles. The aircraft will cruise at approximately 40,000 ft and thus be substantially above cloud layers. The COIL radiates at a wavelength of 1.3 $\mu$ and is being designed to radiate multi-megawatts of energy so that it can heat missile structures to their failing point, causing a destructive kill of the missile.

If the development of the ABL is successful, it will prove to be a great asset to a theater commander-in-chief. It will have an engagement ability to destroy at least 20 enemy missiles. Depending on engagement geometry, atmospheric turbulence, and missile type, it could destroy up to twice that number or more. With an in-air-refueling capability, the range and on-station endurance of a 747 implies that the availability of local in-theater basing will not be a major limitation of the ABL system. If the system performs as the Air Force projects—that is, as part of a tiered theater missile defense architecture operating in concert with various ground-based and sea-based systems—the ABL should provide a flexible, rapidly deployable response for expeditionary operations.

Although the committee believes that the development risks associated with the ABL are reasonable and that they are likely to be resolved successfully, some development risks do exist. They are as follows:

- The packaging and operation of the system on a 747 aircraft,
- Uncertainties related to optical propagation and beam spreading,
- The impact of countermeasures on the system's lethality, and
- The false and missed alarm rates of the ABL's autonomous target detection system.

The Air Force has an active and well-funded effort under way to install and test the laser on a 747 aircraft. An important component of the research and test program deals with propagation and beam spreading, along with the problems associated with holding the beam on the most vulnerable part of the target.

The main operational issue that must be addressed is that the consumables used in the laser cannot be replenished on station. Therefore, any countermeasure that increases the dwell time required for the laser to destroy the target directly decreases the capacity of the system. Coupled with the multiplier effect of the number of aircraft needed to ensure one aircraft continuously on station, this suggests that the mission capacity versus countermeasure trade-off is significant.

Assuming continued funding and no development delays, an airborne laser weapon designed to kill ballistic missiles in their ascent phase is planned for an initial operational capability between 2008 and 2010.

The committee was briefed on the use of a lower powered COIL laser on a low-flying aircraft for defense against cruise missiles. Such a system concept is not funded (nor should it be emphasized) and is not as important to the cruise missile defense problem as ABL is to the ballistic missile defense problem.

## 4.4 BMDO MISSILE DEFENSE R&D PROGRAMS

### 4.4.1 Overview of Theater Ballistic Missile Defense Technology Issues

Technology programs for ballistic missile and cruise missile defense take place in much different programmatic environments. Since the advent of the strategic defense initiative (SDI) in 1983, funding for BMD has been the responsibility of BMDO (originally known as the Strategic Defense Initiative Organization). While the SDI was heavily focused on technology, BMDO currently expends almost all its funding on acquisition and related activities (Figure 4.4). The allocation to BMD technology, exclusive of that earmarked for the space-based laser program, is only 3 percent of the total BMDO budget, far below the 10 to 12 percent generally viewed as the minimum required to prevent technical obsolescence of the major defense acquisition programs (MDAPs) and keep abreast of the advancing threat.

The committee is deeply concerned about the widening gap between available technology resources and the requirements imposed by a missile threat that is growing rapidly in quality as well as numbers. As described below, the gap creates a major issue with respect to the spiral development strategy that is being used for TMD systems development. Moreover, and of equal concern, there is inadequate funding to develop and evaluate more innovative approaches that could significantly improve system cost and/or performance.

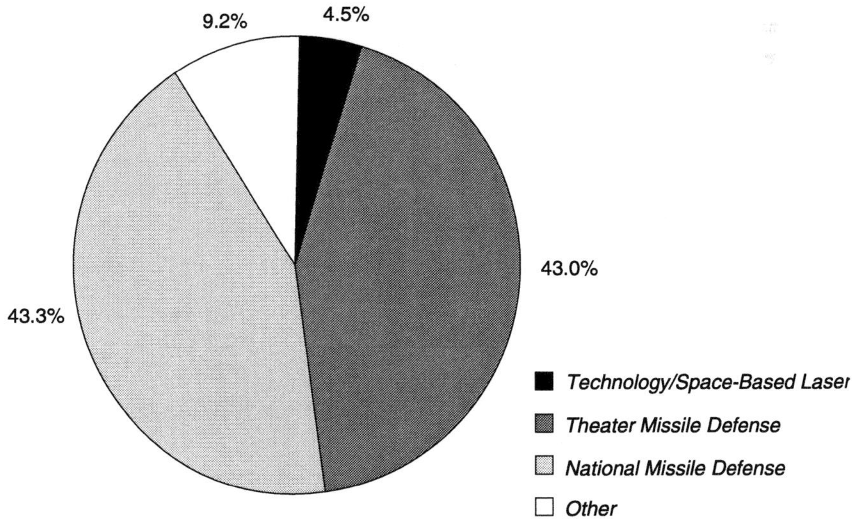


FIGURE 4.4 BMDO FY01 budget ($4.5 billion).

A spiral development strategy has been adopted by BMDO, and all BMD program managers have been directed to formulate plans to implement it. According to BMDO officials, the upper-tier TMD programs—THAAD and NTW—are further along in implementation plans than the lower tier, but all programs suffer from a serious mismatch between programmed technology dollars and spiral objectives. BMDO leadership is said to be keenly aware of this mismatch and committed to increasing the technology budget in FY02 and beyond to bring the enabling technologies along to support spiral objectives.

What the Navy is calling spiral development in the missile defense context is basically just evolutionary development to avoid delaying the initial deployment of BMD systems until all objectives are met. The concept is that an 80 to 90 percent solution should be accepted for initial deployment, with the proviso that a systematic plan for periodic upgrades will be implemented. In a number of briefings on spiral development roadmaps that were presented to the committee, the interval between upgrades was relatively short, and the magnitude of the upgrades was smaller than traditionally has been the case. The thinking behind the strategy appears to plug in performance-enhancing technologies when they become ripe for application rather than to wait for some critical mass upgrade.

It is unclear to what extent the program managers are applying their development budgets to technology upgrades instead of looking to the advanced-technology 6.3 programs to ready the technology for insertion. However, it appears that they are allocating a relatively small proportion of development budgets for technology upgrades and that the projects are highly dependent on 6.3 technology products.

Historically, the process of effecting technology upgrades worked best when the technology program and budget were external to the project. The natural tendency of program managers is to reprogram technology dollars, as necessary, to cover shortfalls in the development program. Because of this, the budgets for some earlier BMD technology programs, such as the Advanced Ballistic Missile Defense Agency, were "fenced" so as to prevent their reallocation to cover near-term exigencies. The committee believes that this general practice is relevant to the implementation of spiral development and further believes that BMDO and the Navy follow it.

No matter what the size of the BMDO technology budget, the technology investment strategy must be improved to apply the available dollars to the highest priority needs more efficiently. In particular, the committee believes that a more explicit correlation should be established between credible countermeasures and the technology solutions to such countermeasures. In too many cases, the technology projects that are being funded bear only a loose relationship to quantifiable improvements in performance against advanced threats. A tighter relationship will clarify the relevance of the funded technologies and will aid in winning increases in the technology budget.

The committee believes that the Department of the Navy should take the initiative in creating an all-Service methodology for relating threat drivers to technology solutions that facilitates an understanding of the critical paths and the commonality of technologies between programs. In this regard, two candidate techniques being used in different parts of the BMD community merit consideration. They are (1) a branch-and-block approach that explicitly networks threat branches with candidate technology responses and (2) breakpoint analysis, which extends the severity of the threat elements until they break the system. Both approaches have merit in explicating the rationale for technology programs and evaluating the payoff of candidate technologies. The committee does not endorse a particular implementation of tools of this type but does recognize the need for an explicit methodology that is common to the principal development programs.

In the following subsections, technology needs and issues are discussed, ongoing programs are assessed, and research priorities for theater ballistic missile defense are suggested.

### 4.4.2 BMDO Missile Defense R&D Programs for Acquisition and Seeker Sensors

Before a cruise or ballistic missile threat can be negated, the threat first must be detected by one or more sensors. Then it must be tracked well enough so that it can be handed over to the defending weapon's seeker soon enough to permit successful negation. Because the typical trajectories of the two missile types are different, the acquisition sensors need to be considered separately.

#### 4.4.2.1 Acquisition Sensor Technology Needs and Issues

For ballistic missile defense, initial detection, which involves discrimination of the weapon-bearing RV from the associated cloud of booster fragments and countermeasures (if any) in order to generate fire-control-quality tracks for handover of the RV target to the kill vehicle, is normally carried out by one or more long-range, ground-based radar sensors.

Prior to this acquisition, other nonradar sensors (DSP, SBIRS-high, and others) may have provided initial threat warning and cueing to speed up the process. However, acquisition, discrimination, and the establishment of a fire-control-quality track on a specific threat RV are the tasks of the radar. CEC, while superficially appearing to be a collection of separate, communicating radars that might be thought of as cueing one another, in fact functions as a single distributed radar that is ideal for tracking air vehicles where line-of-sight issues can arise.

The radar detection challenges presented by ballistic missiles with lofting trajectories lie in the long ranges at which it is hoped to be able to engage them.

The increasingly small RCSs that can be achieved by RV technology (rather than the terrain masking that may be encountered in the cruise missile scenarios) are a further challenge. Increased radar transmitter power can increase range performance. All TBMD radars, including SPY-1 and GBR, include high-power module technologies (e.g., GaAs, GaN, and SiC transmit/receive (T/R) modules) that can increase transmitter power and that are under active development in a number of different science and technology organizations.

Power alone is far from the whole solution. The detection of very small targets requires very low detection thresholds, which inevitably produces a large number of additional "targets" (e.g., false alarms and small pieces of debris) that must be sorted out and distinguished from the actual RV by a discrimination logic. Discrimination of the RV from all the other apparent ballistic objects that may accompany or appear to accompany it is the next step in the threat acquisition process and presents formidable challenges. Clearly the radar must be capable of resolving the individual objects—that is, separating them in all spatial dimensions well enough to allow the RV to be successfully identified and tracked. Since measurement resolution depends primarily on signal bandwidth, high-bandwidth radar is needed.

It seems obvious that the greater the number of individual parameters that can be measured about each candidate RV in the cluster (e.g., length, width, body-motion characteristics, reflectivity, effects on polarization), the better the chance that the real RV can be identified. Consequently, BMD radars need to be capable of multimode operation and must have the ability to make a variety of accurate measurements. Algorithms must be developed to exploit these features. Because of the radar reflection properties (i.e., coherence and isolated scattering effects) of typical objects, increasing bandwidth to gain dimensional resolution finer than that already available in today's X-band BMD radars does not produce more useful images (e.g., more accurate measurements of length).

Since any given radar waveform produces only some but not all the measurements that might be desired, a sequence of different measurements must be carried out. The interesting question then arises of how to optimize the order in which the measurements are attempted. Some BMD radars (e.g., GBR in the THAAD program) address this optimization by considering the radar as an adaptive measurement system, not just a "radar." The waveforms and radar modes utilized at any instant are adaptive responses to the results of the previous measurements. Such adaptive approaches should be explored and extended since they may prove more robust than preplanned approaches in which unexpected RV or countermeasure characteristics are encountered. This approach may also prove useful for the seeker's discrimination tasks as well, with active capabilities expected to be added to the passive sensing currently employed in the exoatmospheric HTK vehicles.

The final step in the acquisition process—that of establishing fire-control-quality tracks on the candidate RV (or RVs, depending on the success of the discrimination process)—is a straightforward and familiar sensor task given the resolution capabilities needed for discrimination. The longer the objects under track can be observed with high resolution and good signal-to-noise, the better the tracking filter estimates will be. In addition, as track quality improves, the volume of the handover basket that is passed on to the HTK vehicle decreases. This in turn minimizes the time required for the onboard seeker to acquire, discriminate, and target the correct RV target. As a result, the kill vehicle divert capability needed to accomplish the intercept will also be minimized.

#### *Needed Acquisition Sensor Research*

Some obvious sensor improvements (e.g., X-band THAAD equivalent ) are not "research" issues but simply need to be done—their pursuit is a priority for the naval forces. Other needed improvements include power and bandwidth improvements via wide-band-gap materials such as GaN and SiC and low-cost T/R modules based on GaAs (X-band) or Si (S-band) microwave/millimeter wave monolithic integrated circuit technology. Digital radar offers flexibility for adaptive measurements—digital waveform generation is often used, but only for limited sets of waveforms (e.g., linear chirps.) Much more flexibility is possible. Any kind of waveform can be generated and processed digitally. The implications of this kind of flexibility should be investigated.

#### *Needed Research on Adaptive Discrimination*

Full digital with analog-to-digital conversion at each element would offer digital phase shifting with no bandwidth limitations when using digital optical communication technology to transfer microwave signals as digital bit streams and cycle-slip plus digital filtering for interpolation to implement the phase shifting. This is not practical at present owing to the performance limitations of current analog-to-digital converters (ADCs) and their expense. There is a need to develop high-performance, inexpensive ADCs and digital receivers.

### 4.4.2.2 Seeker Sensors

The sensor suite onboard the kill vehicle must first acquire the incoming threats or candidate RVs. Missile seekers have an inherently limited field of view and search capabilities. The acquisition of a target (or target complex, as is likely for ballistic targets) requires the designation of a handover "basket" from the long-range sensors. The better this can be accomplished (i.e., the smaller the basket), the better the capabilities of the seeker sensors can be employed to acquire the target as soon as possible. This efficiency, in turn, will maximize the time available for discrimination, aim-point selection, and vehicle end-game maneuvers, thereby minimizing kill-vehicle divert requirements and maximizing the probability of successful intercept.

For ballistic threats, at handover the missile/kill vehicle seeker is typically presented with a many-object target complex depicted and characterized by the measurement capabilities of the initial acquisition radar, from which the target (i.e., RV) must be correctly identified as soon as possible. Clearly, high-resolution "imaging" in all dimensions is required to detect and examine each candidate RV for discrimination. Given the limited aperture imposed by typical missile dimensions, high-frequency systems (i.e., electro-optical and millimeter-wave) must be used. While passive sensors can be adequate for precise azimuth-elevation measurements, for precise range and Doppler measurements, active capabilities via LIDAR or wideband radar adjuncts would be desirable. Current candidate Navy TBMD missile seekers (e.g., SM-2 Block IVA and lightweight exo-atmospheric projectiles) rely entirely on passive optical sensors for the terminal phase, although combined passive/active optical seekers are under development.

Because of the range of possibilities, which includes sophisticated countermeasures, it is clear that the more unique measurements a seeker can make on the totality of objects in the target complex, the better the chance of correctly identifying the real RV. Multiband optical (several IR and possibly visible) sensors with laser detection and ranging (LADAR) and/or millimeter-wave (MMW) radar active adjuncts seem to be called for. If affordable and physically realizable, the combination of multiple optical bands with RF measurements offers good decoy discrimination potential. Although decoys may be produced that are excellent replicas of a RV, the designer of decoys finds it difficult to replicate RV signatures precisely in all-sensing modes.

For many years, BMD discrimination research has concentrated on the development of algorithms derived from observations by a single type of sensor. The more mature discrimination techniques are based on radar measurements, but there is also a significant body of work on passive optical sensor discrimination. Only recently has serious attention begun to be applied to fusing of radar and optical data to enhance discrimination performance.

The main BMD systems currently under development all have both radar and optical sensors, and the enhanced discrimination potential achieved by combining the data they collect on various features of the threat objects is becoming increasingly evident. The radar data, particularly that measurable by the X-band radars being developed, allow the precision measurement of microdynamic features of threat objects. The passive IR sensors being developed for performing onboard interceptor functions are naturally adept at measuring thermal characteristics of threat objects. In addition, there is a large class of features, such as macrodynamic body motions, that both sensors can measure. The potential for a significant improvement in discrimination capability lies in the effective fusion of these feature vectors.

While passive sensors can make simultaneous measurements (e.g., using FPAs) on multiple objects in multiple modes, active sensors typically employ different waveforms sequentially. Here, as for long-range sensors, treating the seeker's active component (and perhaps the computer resource) as a measurement system rather than simply a radar or an imaging system, may permit effective adaptive procedures to be applied.

### *Needed Seeker Sensor Research*

The following seeker sensor research is needed:

- Multiband IR/visible sensors with laser radar or radar adjunct should be developed to address the discrimination issues that are certain to arise as ballistic missiles continue to become more sophisticated. If one looks at the dramatic advances in focal plane materials and mechanization technologies, it is easy to project continued improvements in quantum efficiency, sensitivity, bias and noise suppression, and resolution. The use of additional resolution and narrower detector bandwidths opens the possibility of other multispectral discriminants, including materials and imaging.
- Discrimination algorithms that exploit all the signatures that can be detected by multispectral sensors should be developed.
- LIDAR systems with multipixel FPA, which measure range-to-pixel, need to be developed in order to avoid the mechanical complexity associated with scanning optical systems.
- More powerful lasers are probably needed to extend the range of the three-dimensional imaging LIDAR adjuncts.
- RF/MMW adjunct possibilities for enhanced discrimination should be explored, including the possibility of deployable antennas for exo-atmospheric intercepts, to mitigate the limitations of kill-vehicle dimensions.
- Adaptive discrimination algorithms using the active capability of the seeker as a measurement tool need to be developed.
- Multiband LWIR sensors and their associated algorithms are able to reject most celestial objects and background by temperature and/or lack of movement; however, visible light sensors must deal with this problem. Background obscuration algorithms should be explored to deal with low-cross-section targets against the stellar background, although because the RVs are so small and the sky background is so complex, this approach seems to be less promising than it was for cruise missile threats.

### 4.4.3 BMDO Missile Defense R&D Programs for Weapons

#### 4.4.3.1 Interceptor Technology Issues and Needs

The principal goal of TBMD weapons research is to assure that future interceptors can engage incoming missiles with a high probability of success. In this context, success means preventing missile warheads—including weapons of mass destruction—from destroying their intended targets. A need exists to cope with increasingly sophisticated missiles that could be launched in closely coordinated salvos at one or more protected assets. Future missiles may be more difficult to detect, may contain multiple warheads, and may be capable of maneuvering at any point on their trajectory. A field of false targets may surround the warheads that are delivered by future missiles.

Secondary goals of TBMD research include the achievement of high reliability at the component and system levels, high effectiveness, and low per-missile and total system life-cycle costs. Tertiary goals are the development of backup strategies and alternatives in the event of program failures, increased onboard autonomy, improved performance within existing missile magazine constraints, improved methods for empirical test and evaluation of total effectiveness, and refined algorithms for computer-based analysis and design.

Because the missions and interceptors are different for NAD and NTW interceptors, specific needs and issues are different as well. Each TW interceptor is entirely dependent on the satisfactory performance of a single kinetic-kill vehicle (KKV). Since the KKV does not contain an explosive warhead, it must achieve a very small miss distance with respect to its intended aim point on the RV to achieve a successful intercept. There is a four-dimensional set of requirements for KKV performance: discrimination, accuracy, response time, and adaptation. Once the KKV has been delivered to its operational basket with an orientation that places the target in the fields of view of the onboard sensors, it must discriminate the correct target from a field of targets. If the correct target is maneuvering, the accuracy and response time of the KKV's guidance-and-control system must be good enough to ensure collision with the target. There is uncertainty in the predicted maneuvering intercept point and a need to converge the actual miss distance to a sufficiently small value. In statistical terms, not only the mean but also the covariance of the intercept-point error must be close to zero. The margin for error is relaxed considerably if the KKV cross section can be increased (by, for instance, expanding the structural cross section) or if multiple KKVs can be carried by the interceptor, increasing the probability of a hit.

This same four-dimensional set of criteria applies for the NAD interceptor. Even if a NAD interceptor is guided to zero-miss distance, its exploding warhead will considerably expand its effective cross section, and the warhead explosive and fragments provide added effect. The likely normal load factors ($g$ levels) for both the maneuvering incoming missile and the interceptor are considerably higher in the atmosphere, so guidance-and-control responsiveness is increased. For TBM intercept, the time available to discriminate, close, and hit is shorter than for the exo-atmospheric case. However, the atmosphere helps the defense to filter out lightweight, nonthreatening elements.

The overlapping cross sections of the interceptor and its target define the hit-to-kill lethal radius. As illustrated in Figure 4.5, when all factors are considered, the overall probability of a successful intercept for any given TBMD system increases as the lethal radius increases. The probability is small when the lethal radius is below some critical value, and it is relatively constant above the "knee" in the curve. The goal of any research and development program should be to move the knee as far to the left as possible.

For 30 years, beginning in the mid-1960s, the Navy maintained a high-energy laser program that was oriented toward the development of a shipboard high-energy laser (HEL) that could be used for ASCMD. The result of this extended investment was a realization that in a moisture-laden maritime environment, laser fluences were reduced by scattering and absorption to a point where they would have limited usefulness as a ship self-defense weapon.

More recent appraisals of the value of HEL as an ASCMD weapon have centered on the observation that an HEL can char the nose cone of an incoming missile. If the missile is radar-guided, the charring will result in increased transmission loss through the nose cone and a partial or complete loss of radar guidance. If a radar-guided missile loses its radar capability, it is generally programmed to fly in a straight line to its target based on its IMU.

A missile that flies in a straight line toward its target is more vulnerable to destruction by short-range defensive weapons than a missile that makes evasive

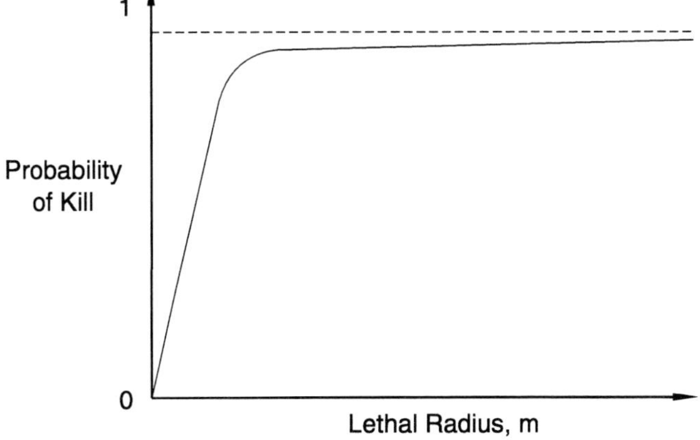


FIGURE 4.5 Probability of kill vs. lethal radius of intercept (notional).

maneuvers. Thus, some have argued that a shipboard HEL should be thought of as a complementary weapon that might increase the probability of kill of associated short-range defensive weapons. Others point out that an ability to continue evasive maneuvers based on IMU guidance can be (and in fact has been) programmed into modern missiles as a backup mode of operation in the event that their radar is jammed. Unless unforeseen changes occur in HEL technology, the committee doubts that shipboard HELs will be introduced prior to 2020.

Some directed-energy weapons, in particular lasers, were considered as weapons for the destruction of incoming missiles during late 1980s in the context of the SDI, which envisaged space- as well as land-basing of these systems. Lasers have the potential to deliver a lethal dosage of energy at a great distance with minimal delivery time (velocity of light).

Lasers in the theater ballistic missile defense arena have a role to play principally in the boost phase, when the target size is the largest because the booster is still an integral part of the target. Moreover, the booster is perhaps the softest component of the target. Once the warhead has separated from the propulsion vehicle, the target's size, velocity, and hardness all work together to make the problems for laser weapons a lot more difficult. In the midcourse phase of flight, the opportunities for simulation/antisimulation and decoys abound, making discrimination of the actual warhead difficult. For boost-phase intercept, depending upon the primary boost vehicle, intercept must take place below separation altitude, which is typically less than 100 km. This, in turn, requires the laser platform to be in position to track and engage the target with enough fluence and integration time to create the damage. While the energy is delivered at the speed of light, the rate at which it arrives and is absorbed by the target requires illumination times of several seconds depending on the construction, material, and surface finish of the target.

An extended discussion of the Air Force's laser weapon program is provided in a separate section. The program is based on COIL technology. Another class of lasers, the FEL, attracted much attention during the SDI era. Interest in the FEL appears to be making a comeback because of the wavelength tunability feature, which allows the selective use of a wavelength that has minimal absorption during its transmission through the atmosphere.

The Department of Energy's Thomas Jefferson National Laboratory has been working on a system that appears to be scaleable to the multimegawatt levels of laser power needed for theater missile defense applications. Because of their size and electrical energy requirements, as well as the need for shielding the radiation from the beam dump, FELs would only be suitable for ship (and perhaps also ground-based) deployment. These lasers were proposed by the Los Alamos and Lawrence Livermore National Laboratories and by Boeing and others during SDI development as ground-based lasers that, in conjunction with space-based optics, could be used in ballistic missile defense applications. However, the laser power requirements for such applications were seen to be extraordinarily high.

In spite of their promise, FELs are still very much laboratory devices. Their deployment in practical defense systems is farther out in time than the possible deployment of COIL and hydrogen fluoride/deuterium fluoride lasers.

#### 4.4.3.2 Interceptor Technology Programs

Two related programs are the BMDO-sponsored AIT and exo-atmospheric interceptor technology (EIT) programs, led by the Army and Navy, respectively. Key technologies for AIT include low-cost, high-performance strap-down seekers; lightweight and highly reliable solid-propellant, divert and attitude control systems; and accurate modeling of jet interaction for hit-to-kill intercept in the atmosphere. The EIT focus is on the development and demonstration of two-color infrared sensors combined with an active LIDAR sensor and lightweight composite materials for shrouding, construction, and component housing.

#### 4.4.3.3 Technology Priorities

Although the committee did not conduct an in-depth analysis of BMD interceptor technology programs, it believes that these programs are addressing the right issues. However, the committee questions the BMDO technology program as a whole—that is, whether it is funded to a level that realistically supports the BMDO spiral development strategy. In developing an enhanced program, the following issues should be considered:

- Exo- and endo-atmospheric interception of maneuvering targets;
- Enhancements of, or alternatives to, hit-to-kill strategies; and
- Improvement of algorithms for multidimensional state vector estimation, and for prediction and control.

The committee notes, in addition to inadequate support of spiral development, the lack of high-risk, high-payoff interceptor technologies in the BMDO technology program. Concepts such as multiple, highly minaturized kill vehicles per interceptor need to be under development to provide a counter to future advanced threats and to remedy shortfalls in discrimination and lethality.

The committee endorses the development of a 21-in.-diameter second stage for the SM-3 interceptor. Some of the suggested growth paths such as two-color cooled optics interceptor sensors, increased divert-and-maneuver capability, and kill enhancement mechanisms all drive the kill-vehicle weight higher, which in turn requires more boost energy.

### 4.4.4 BMDO and DARPA BMC3 Efforts

Technology issues and needs for BMD BMC3 are discussed above in Section 4.1.3.1. Here, directly relevant BMDO and Defense Advanced Research Projects Agency (DARPA) BMC3 efforts are briefly discussed. Not much work is being done on BMC3 technology for TMD, although there are technology projects toward BMC3 for NMD.

The battle management command, control, communications, computing, and intelligence (BMC4I) advanced technology (BAT) program is developing BMC4I technologies for NMD and TAMD. Technical objectives include improvement of kill assessment processes, communications links with interceptors, high-performance computers, and engagement planning algorithms. Technology for NMD capability 1 includes upgraded early warning radar (UEWR) tracking and fusion algorithms. Technology for NMD capability 2 includes tracking of multiple, closely spaced objects; discrimination algorithms; target engagement and fire control; and kill assessment techniques.

BMDO created project Hercules in January 2000 as a National effort to develop robust, adaptive algorithms to counter off-nominal and evolving threats. The project incorporates a spiral development process and aims to develop and transfer algorithms to the MDAPs. It follows a design-to-capability approach, meaning that the algorithms being developed are associated with threat parameters and contain allowances for performance margins. Its primary objective is to develop algorithms that are less dependent on a priori threat data than is currently the case.

Project Hercules addresses the critical BMD functions of tracking, discrimination, aim-point selection, and kill assessment, as well as a number of BMC3 functions. A high priority is assigned to discrimination algorithms. The very close relationship that has been established with the other Services is largely fostered by four teams reporting to the project office. In addition, a large number of companies, universities, and federally funded research and development center agencies are participating in the program. A number of specific deliverables to the participating MDAPs have been identified and scheduled by the project.

It should be noted that a modest research investment is being made at DARPA in research on how to structure such dynamically assembled systems. Most of these efforts are primarily concerned with agent-based systems (e.g., the control of agent-based systems (COABS) program in DARPA's Information Systems Office (ISO) and the autonomous negotiating teams (ANTS) program in DARPA's Information Technology Office (ITO)). Other efforts at DARPA's ITO are concerned with the use of distributed networks of sensors (e.g., the sensor information technology (SENSIT) program) and with assembling many independent vehicles into a single coherent sensor or effector. While none of these programs have a specific focus on BMD, they do show that DARPA is interested in examining issues relevant to BMD and offer a possible avenue for cross-Service collaboration in early experimentation.

## 4.5 INTERNATIONAL PROGRAMS

### 4.5.1 United States-Japan Cooperative Program

In August 1999 the United States and Japan signed a memorandum of agreement to perform cooperative research and development on a ballistic missile defense system. This R&D effort shows promise of a disciplined approach to the system engineering of a capability that would defend Japan from continental threats and potentially serve as the Block II NTW system. The system is scheduled to achieve an initial operational capability sometime after 2015.

Task 1 of the effort, currently under way, focuses on conceptual definition. It has delivered or will deliver a threat description, operational system guidelines, a scenario reference mission, technical assessment reports, and a preliminary integrated Dem/Val phase plan. Trade-off studies are being conducted to define the system as a whole and especially its missile element. In particular, cooperative design studies are addressing the missile's guidance unit, its divert and attitude control system, second-stage propulsion, and a lightweight nose cone.

Planned follow-on tasks will result in a preliminary design of the Dem/Val missile (in a prime item development specification) and perform risk reduction demonstrations.

### 4.5.2 United States-Israel

Chemical hydrogen fluoride/deuterium fluoride (HF/DF) lasers, which at one time were considered key candidates for ground-based lasers with space-based optical components for SDI, have been reconfigured as a tactical high-energy laser (THEL). In a cooperative activity with Israel, this system has been operated at a power level of 300 to 400 kW and has shown an ability to destroy Katyusha rockets in flight at a standoff distance of about 1 km (or less).

Such lasers—which because of their large size and long wavelengths force the use of large optics for targets and long distances—may be candidates for deployment on ships for the defense of Navy assets against cruise missiles and enemy aircraft attacks. They would be in competition with tunable FEL lasers, whose development has shown promise in recent years. Between about 1965 and 1985, the Navy sponsored an extensive program of research in HF/DF lasers. Enthusiasm for shipboard lasers waned as a result of atmospheric adsorption, beam scattering in the maritime atmosphere, and the realization that modern missile nose cones could sustain high levels of damage.

The THEL facility is housed in a building at White Sands' High Energy Laser Systems Test Facility. The beam comes from a small turret on top of the structure, attached to which is what amounts to a battlefield chemistry project: tanks of chemicals that, when mixed, generate enough energy to fire the laser. This facility is likely to remain in White Sands, New Mexico, where it will serve as a research platform, rather than to defend Israel's northern border, as originally intended (although it still might go to Israel if that nation wants it for an emergency). U.S. and Israeli military authorities would like to develop a smaller, mobile version of the THEL demonstrator at White Sands. Both the Israeli Defense Force and the U.S. Army are interested in fielding a short-range battlefield defensive laser system that would be able to shoot down artillery rockets, mortar shells, and—possibly—aircraft and cruise missiles. The development of a mobile THEL has been estimated to require 5 to 7 years of additional effort.

In tests starting in June 2000, it was fired at 16 rockets and one winged insect that landed on the beam emitter at precisely the wrong time. Some of those rockets were fired simultaneously to test THEL's ability to engage multiple incoming targets. The demonstrator was built with combat capability in mind, and officials had discussed moving it from White Sands to Israel. Despite the successful tests of 2000, there has been a growing concern that the demonstrator was not ready for action. Israel has also expressed concerns that a fixed THEL would become a difficult-to-defend target for attacks. The changing situation in Israel has made a fixed THEL less acceptable. During the summer of 2000, Israel abandoned its occupied territory in southern Lebanon, allowing Hezbollah to launch attacks much closer to the Israeli border. More THELs, or a mobile system, would now be needed to properly defend the border area.

The U.S. and Israeli governments are reported to be completing an agreement that would provide for the development of a mobile THEL. The contemplated mobile system would be carried on a few heavy trucks (one truck would have the laser, another would have the rocket-tracking radar, and others would carry the fuel chemicals). The design objective would be to produce a system light enough to be transported by a C-130 cargo plane.

The committee believes that the Department of the Navy should certainly continue to track the progress of the U.S.-Israeli THEL and consider it for possible application in a maritime environment.

The Israeli Arrow Program is a cooperative program funded largely with BMDO money. It is designed to engage both endo-atmospherically and high in the atmosphere, much like THAAD. While quite large, Arrow uses several technologies similar to those being pursued by the Navy systems and will yield useful information on fragment warhead lethality and seeker phenomenology.

# 5

# Conclusions and Recommendations

## 5.1 CONCLUSIONS

ASCMD, OCMD, and TBMD capabilities are essential if naval forces are to operate in littoral areas. Threats to naval and joint forces operating in littoral areas stress the capabilities of current ASCMD, OCMD, and TBMD systems, and all indications are that they will become even more stressing. Future cruise missile threats are likely to be characterized by features such as low-altitude terminal flight paths, low-RCS values, high agility and Mach number, ECM-resistant sensors, and precision terminal homing capabilities.

The Marine Corps operational concepts embodied in OMFTS and STOM will not be feasible without effective TBMD and OCMD if operations are conducted in this threat environment. In addition, the requirements to provide naval surface fire support, OCMD, and TBMD in littoral areas will mean that its ships must operate in near-shore waters, where their survival will be totally dependent on the availability of robust ASCMD capabilities. Current ASCMD systems have marginal or poor performance in littoral areas against some existing advanced ASCM threats. The Navy has many significant improvements under development—e.g., MFR, SSDS, ESSM, SPY-1D(V) radar—which should be fielded as soon as possible. Some needed components are not under development (e.g., an ESSM launcher for non-Aegis combatants). Furthermore, naval combatants need an elevated detection platform and an over-the-horizon engagement system to restore an area defense capability providing the depth of fire needed for robust defense.

Unless there are significant (and unanticipated) increases in Navy budgets that would permit new classes of sensors to be fielded, the negation of OCM and ASCM threats will require the Navy and the Marine Corps to field new sensor and weapon capabilities and/or to become dependent on and integrated with the nonorganic sensor systems of other Services and agencies. These nonorganic sensor systems might include AWACS, JLENS, UAVs, and DSP's SBIRS-high and SBIRS-low. If budget resources were to become available, any new sensors that might be fielded by the Navy and Marine Corps should include an elevated AMTI capability consisting of either a suitable radar (hosted on an E-2C or other airframe) or a multistatic system based on UAV receivers and AWACS, E-2C, or JLENS transmitters.

Future ballistic missiles are likely to be characterized by features such as spin-stabilized RVs, separating ACMs and RVs, low-observable RVs, maneuvering and tumbling RVs, and an ensemble of penetration aids that might include decoys, shrouds, jammers, and debris.

The NAD and NTW Block I systems will enable defeating some current unsophisticated ballistic missile threats; however, until upgraded systems are fielded, these systems will have limited capabilities against postulated advanced ballistic missile threats. Nonseparating theater ballistic missiles can be engaged and negated by these systems, although hitting the warhead of a tumbling vehicle remains a challenge. However, NTW Block IA and B will not be capable of providing simultaneous TBMD and ASCMD/OCMD. NTW Block IA and B ships will require the presence of supporting ships. Although NTW Block IC will integrate TBMD with other Aegis capabilities, NAD and NTW Block IC will not provide a robust capability for negating ballistic missiles with sophisticated penetration aids. This will require a substantial increase in radar capability over the SPY-1.

The SM-2 Block IVA and SM-3 programs appear to be well structured, but upgrades are required to the SPY-1 radar to make its capabilities compatible with the reach of the SM-3. These might include increased propulsion for the SM-3 to provide better performance robustness in the face of payload growth uncertainties and an improved HTK vehicle characterized by enhanced divert capabilities, two- or three-color IR sensors, laser radar, and so on.

Although both the NAD and NTW systems are based on the concept of spiral development (build-improve-build-improve . . .), the R&D to support such a development concept is not in place. BMDO controls investments in missile defense R&D, and its current levels of such investment are insufficient to result in significantly improved capabilities. BMDO control of missile defense R&D is a double-edged sword. On the one hand, it discourages Service investment. On the other hand, it fences missile defense R&D from other Service priorities. Service-sponsored R&D in support of out-of-the-box solutions for missile defense can only take place in special access programs that often do not result in the acquisition of operational systems.

Naval forces lack a competent battle management command, control, and communications (BMC3) capability in terms of both concepts of operation and systems for missile defense in coordination with offense operations in littoral areas. Inadequate procedures and technical capabilities exist for coordinating assets in the battle space, and current enhancement efforts are often based on legacy technology (e.g., Link 16) that does not support the necessary flexible modes of operation. Link 16 and CEC cannot be evolved far enough to provide the necessary flexible connectivity. Evolving commercial wireless technology can be leveraged to meet communications needs—thereby disentangling the communication problem from the battle space coordination problem. Current efforts to improve battle space coordination must be continued (and possibly augmented).

The Navy and Marine Corps have no current capability for OCMD and no accepted CONOPS for it. The CONOPS for NTW and NAD are still evolving. Neither naval nor joint operations are addressed in a comprehensive way. Commanders in chief (CINCs) generally must create their own CONOPS based on ad hoc situations that arise in their area of responsibility. The Joint Forces Command has overall responsibility for development of CONOPS for theater missile defense.

Airspace deconfliction while conducting effective TMD is an important and difficult problem that does not appear to have been addressed adequately.

## 5.2 RECOMMENDATIONS

During the course of this study the committee developed recommendations to address the following issues:

- Prioritization of cruise and ballistic missile defense programs;
- Stove-piped theater missile defense systems;
- Limitations related to the concept of operation for the conduct of OCMD and TBMD in the course of expeditionary warfare operations when joint and coalition forces are present;
- ASCMD, OCMD, and TBMD deficiencies and the programs to correct them;
- Current and projected Marine Corps OCMD capabilities;
- BMC3; and
- Technology investment.

The committee's recommendations are discussed next.

### 5.2.1 Prioritization of Cruise and Ballistic Missile Defense Programs

Antiship cruise missile defense, overland cruise missile defense, and ballistic missile defense will all be necessary for naval (and joint) forces conducting 21st-century military operations for a number of reasons:

• ASCMD—Antiship cruise missiles in the hands of potential adversaries are numerous, sophisticated, and widespread. Every naval combatant becomes a target whenever it enters a theater and must defend itself effectively if it is to be an asset rather than a liability.

• OCMD—In the future, land attack cruise missiles will allow potential adversaries to deny military forces access to ports, airfields, and other entry points. In effect, the Navy has no OCMD capabilities, and building such capabilities will require time and investment.

• BMD—Tactical ballistic missiles are widespread weapons of terror and potential mass destruction. Naval forces need capabilities to provide ballistic missile defense to ports, airfields, and other entry points until assets arrive in-theater from other Services. In the future, longer-range ballistic missiles will become more prevalent and an adequate theater ballistic missile defense will require defense in depth.

With the exception of developing a robust capability for OCMD, there is little disagreement within OPNAV and the Navy acquisition community concerning missile defense programs. Moreover, all Navy ballistic missile defense programs are matched to funding limitations or BMDO-imposed cost constraints and as a result have adopted evolutionary development programs that defer the development of necessary capabilities until far into the future.[1]

In the likely event that budget levels will not be sufficient to fund all cruise and ballistic missile defense efforts fully, the committee believes that the Department of the Navy will need to assign funding priorities for R&D efforts as follows:

1. ASCMD,
2. Area defense of forces and assets ashore against both overland cruise missiles and ballistic missiles (NAD system), and
3. The NTW system.

The committee's rationale for according first priority for R&D funding to ASCMD is that if the Navy does not have a robust ASCMD capability, its abilities to undertake or support operations in littoral areas will be seriously limited.

[1]The committee is also concerned that in those areas where naval R&D needs and priorities are not supported by BMDO investment, there is no safe mechanism for the Department of the Navy to apply funding of its own. Furthermore, the committee believes that if the Department of the Navy allocates R&D funds for theater missile defense, congressional committees will most likely cut those funds on the basis that missile defense R&D has already been accounted for in the BMDO budget. In the end, there is no investment for theater missile defense R&D. Therefore, the committee believes that a stronger organizational link should be established between the Department of the Navy, the Office of the Secretary of Defense, and BMDO in order that R&D be supported.

The committee could not come to a consensus on the relative prioritization of R&D funding between OCMD and NAD. All members of the committee recognized that defense against land attack cruise missiles and defense against ballistic missiles are necessary components of the same mission, particularly if the Navy is to protect forces and assets ashore.

Some argued that since ballistic missiles are widely available to probable or potential adversaries, and since land attack cruise missiles currently are not widely proliferated, priority for R&D funds should be assigned to the NAD program. Furthermore, ballistic missiles, which may be configured to carry weapons of mass destruction, can have a major political impact on allies and on forces ashore.

Others argued that the development of an OCMD capability (be it naval or joint) was essential for the protection of forces ashore against a threat that would have a high probability of proliferating if no such defense were to be developed. Those who supported a relatively high priority for R&D funding for OCMD also pointed out that that the most effective means of developing an OCMD capability is through the use of an elevated detection platform. The same elevated platform and sensor system that is needed for OCMD can be used to extend the detection horizons of a surface ship. Thus, sensor developments that will be necessary to provide OCMD capabilities will also contribute to the improvement of the Navy's ASCMD capabilities.

Although the committee could not achieve a consensus on the relative priority for R&D funding between OCMD and NAD, it had significant concerns that R&D funding for the development of a competent OCMD capability has been relatively limited. Unless R&D funding for OCMD is given higher priority than it currently has, the prognosis for the development of OCMD capabilities will continue to be bleak. Furthermore, if the Navy cannot provide OCMD in support of Marine Corps or Army forces ashore, at least in the early stages of operations, then the full potential of naval expeditionary forces (as envisaged in Forward...From the Sea and OMFTS) will not be achieved.[2] Thus, without a land attack cruise missile defense capability to supplement its ballistic missile defense capabilities, the ability of naval forces to influence events ashore will be limited to attacks on stationary targets with standoff missiles and air-delivered ordnance.

In its assessment of the Navy's existing and planned ballistic missile defense capability, the committee would emphasize the NAD system over the NTW

---

[2]Some might argue that in a developed theater the Army's Patriot advanced capability (PAC-3) would be deployed. As currently configured, PAC-3 does not depend on the availability of an elevated air moving target indication radar to detect and track missiles that make maximum use of terrain obscuration in order to evade detection by ground-based radars. Thus, until PAC-3 is provided with a robust capability to negate missiles that employ terrain-obscured trajectories, no OCMD capability exists.

system.[3] The basis for this emphasis on NAD relates to BMDO's role in defense-related development and acquisition for TMD systems. In some developed theaters, competent land-based theater missile defense systems might be predeployed. For example, if the development and deployment of the Army's THAAD system were successful, it could provide significant midcourse engagement capabilities in a theater where it had been deployed prior to the onset of conflict. In addition, if the development and deployment of the Air Force's ABL system were successful, it could provide ascent-phase engagement capabilities against shorter-range ballistic missiles. In such circumstances, the NTW system would supplement the projected capabilities of these systems in addition to the projected endo-atmospheric ballistic missile engagement capabilities of both the NAD and the Army's PAC-3 systems.

**Recommendation: The Secretary of the Navy, the Chief of Naval Operations (CNO), and the Commandant of the Marine Corps (CMC) should assign R&D funding priority in the following order: (1) antiship cruise missile defense, (2) area defenses against both overland cruise missiles and ballistic missiles (NAD system) for the protection of forces and assets ashore, and (3) the NTW system.**

### 5.2.2 Stove-Piped Theater Missile Defense Systems

The committee recognizes that the distributed architectures envisioned for future theater missile defense operations, driven by the realities of the availability and the readiness of defense elements, make it a risky and uncertain business to provide the required level of protection against threatening ballistic and cruise missiles. A significant part of the uncertainty associated with connecting available sensors and shooters into an effective defense network comes from the fact that TMD systems are developed and tested largely as vertically integrated defense systems (as, for instance, are NAD and PAC-3) and are relatively loosely integrated as a family of systems. This suggests that if dynamically assembled distributed architectures are to function effectively, a new paradigm for development and testing needs to be applied by BMDO and the Services.

The committee is well aware that a number of concepts, studies, and research activities have addressed this general issue, with proposals ranging from tightened integration of the family of systems (FOS) to creation of a single distributed TMD system, to replace the concept of an FOS. The Navy's concept for moving from platform-centric to network-centric forces is related to the more narrow issue of TMD addressed here. The scope of such a transformation is

[3] Program budget decision 224 calls for a shift of $121 million from the Navy theater wide program to the Navy area defense program over FY02 and FY03 (*Inside the Pentagon*, January 18, 2001, pp. 12-13).

beyond the charter of this study, but the committee believes that the challenge of utilizing available missile defense resources in a dynamic theater environment needs to be acknowledged and that some of the changes required in the development and testing process to meet that challenge need to be faced.

The process of dynamically assembling defense resources in-theater, otherwise known as network-centric operations, requires that advanced engagement modes be employed, such as "launch on remote" and "forward pass." Some of these engagement modes are beginning to be tested in joint experiments, and the committee strongly believes that such tests should be continued and expanded. Fundamental to the success of network-centric operations, or any other version of distributed architectures, is the ability to break the bonds of the vertically integrated TMD systems without unduly compromising the effectiveness of engagements.

All the TMD systems are being developed and tested with the tight bonds in place, having been trained, in effect, to function as autonomous systems. There is nothing wrong with this classical approach to development and testing, which has been conditioned by the interoperability, layered defense, and advanced engagement mode development and testing that is already in progress. However, when integrated systems are not available to provide defense, the different elements of the systems need to be adaptable for use in distributed architectures in order to enable distributed engagement modes, which need to be thoroughly developed and tested to demonstrate their feasibility.

**Recommendation: The Secretary of the Navy, the CNO, and the CMC should support the expansion of distributed defense development and test plans by the Ballistic Missile Defense Organization (BMDO) and experiments to demonstrate the related advanced engagement modes. To the extent practicable, the system integrated tests being planned by BMDO and experimental programs such as the Theater Missile Defense Critical Measurements Program should be structured and extended to incorporate the critical defense functions unique to distributed architectures.**

### 5.2.3 Limitations Related to the Concept of Operations for the Conduct of OCMD and TBMD in the Course of Expeditionary Warfare Operations When Joint and Coalition Forces Are Present

The Navy has declared expeditionary warfare that will influence events ashore as one of its main missions. The Marine Corps Expeditionary Maneuver Warfare 21[4] (EMW 21) strategy is consistent with and dependent on the Navy's

[4]Expeditionary Maneuver Warfare 21 is the Marine Corps overarching strategy for conducting 21st-century Marine Corps operations, such as those described in "Operational Maneuver From the Sea"; "Ship to Objective Maneuver" (Van Riper, LtGen Paul K., USMC, 1997, "Ship to Objective

capability in this area. Expeditionary warfare and theater missile defense are thus mutually dependent.

Expeditionary operations envision the possibility of forcible seaborne entry into a theater in which Marine Corps forces are launched from Navy ships and proceed directly to targets beyond the shoreline. Such operations may include peace enforcement, noncombatant evacuations, or combat operations. For Marine Corps forces to have the required reach, it will be necessary that ship formations approach the shoreline as needed to deliver supporting fire and logistical support to the Marines ashore. The same kind of support could be required if Army elements are involved as part of a joint task force. In any scenario in the littorals, the Navy must be able to defend both its own ships and the assigned forces against attacks by ballistic and cruise missiles.

Almost all the presentations to the committee began by showing the complexity of coordinating defensive measures with offensive operations in the same battle space. However, it is clear that no consistent CONOPS exists for integrating conventional supporting arms (attack and other helicopters, artillery, naval fires, and close air support) in the offense with TBM and cruise missile defense. The task is even more complicated in a joint operation because of the need for airspace deconfliction. The committee believes that the Navy's ability to defend a carrier battle group or an amphibious task force against aircraft at sea or in the littorals is well established. Likewise, the Marine Corps has clearly defined procedures and CONOPS for both helicopter-borne and surface assaults on objectives ashore as well as a reasonable defensive capability against enemy aircraft and helicopters.

The evolving threat of theater ballistic missiles and cruise missiles presents a challenge too tough for any one Service to counter effectively on its own. While the Services and the war-fighting CINCs accept the need for a fully joint solution to missile defense, progress is slow because each is pursuing a different approach. The committee believes that neither the Navy nor the Marine Corps, which must rely on the Navy for protection at least in the early stages of an operation, will carry out missions in this threat environment without assistance from other joint assets. Therefore, it is imperative that the Navy identify those intelligence, surveillance and reconnaissance assets in other Services that can help it to carry out its TBM and cruise missile defense mission in the littorals. In addition,

---

Maneuver," Marine Corps Combat Development Command, Quantico, Va., July 25, available online at <http://192.156.75.102/stom.htm>); "Maritime Prepositioning Force 2010 and Beyond" (Krulak, Gen C.C., USMC, 1997, "Marine Prepositioning Force 2010 and Beyond," Headquarters, U.S. Marine Corps, Washington, D.C., December 30, available online at <http://192.156.75.102/mpf.htm>); "Sustained Operations Ashore" (Krulak, Gen C.C., USMC, 1998, "The Marine Air Ground Task Force in Sustained Operations Ashore," U.S. Marine Corps, Washington, D.C., June 10, available online at <http://192.156.75.102/soa.htm>); and "Other Expeditionary Operations" (Warfighting Requirements Division, to be published, "Other Expeditionary Operations, Draft Concept Paper," Marine Corps Combat Development Command, Quantico, Va.).

the capabilities of such assets (AWACS, JSTARS, JLENS, and UAVs) should be integrated into a Navy CONOPS that supports the joint force commander under the joint theater air missile defense 2010 operational concept and its developing operational and systems architectures. Testing and demonstrations under this concept should be coordinated with the CINC, U.S. Joint Force Command.

The committee believes that as the Department of the Navy continues to move forward with naval operations concentrated in littoral areas, there will be several operational implications for the integration and coordination of expeditionary and strike warfare assets. It further believes that the Department of the Navy should account for these implications either by accepting the need for changes to concepts of operations or by investing, as necessary, to achieve the technical advances necessary to make preferred concepts feasible. Of immediate concern is the need to achieve a CMD capability to support naval forces and joint forces operating in littoral areas.

**Recommendation: To achieve a competent cruise missile defense capability for the support of naval and joint forces operating in littoral areas, the CNO and the CMC should do the following:**

- **Develop a concept of operations with the other Services that routinely substitutes and employs assets such as the airborne warning and control system (AWACS) air moving target indication (AMTI) radar or the joint land attack cruise missile defense elevated netted sensors (JLENS) to perform over-the-horizon target acquisition and missile command functions envisaged for the E-2C Radar Modernization Program (RMP) radar; and**
- **Leverage joint experimentation in order to develop the operational concepts and technical capabilities necessary for joint missile defense operations.**

### 5.2.4 ASCMD, OCMD, and TBMD Deficiencies and the Programs to Correct Them

Over the past several years, lower levels of R&D investment have allowed the ASCM threat to evolve somewhat more rapidly than shipboard defenses have been improved. Future threats, which are projected to have much smaller radar signatures, greater agility, and electronic countermeasure (ECM)-resistant sensors, may well overstress these defenses when the Navy is constrained to operate in a littoral environment. The proposed acquisition and deployment of SPY-3 and the X-band horizon search MFR, along with some advances in the Navy's electronic warfare techniques, should redress some but not all of the Navy's projected ASCMD deficiencies. The committee is concerned that there are no programs in place to develop additional techniques to increase the Navy's ASCMD effectiveness.

In the final analysis, the ASCMD problem relates to the fact that a low-altitude cruise missile can get relatively close to a surface ship before it crosses the radar horizon of the ship's defensive sensors. If the number of incoming cruise missiles is sufficiently large, their agility and speed sufficiently high, and their RCS sufficiently low, the defensive system will be overwhelmed. A strong layer of short-range self-defense is needed, but robust defense requires a depth of fire that can be provided only by employing elevated sensors, such as the JLENS, that extend the horizon of the defensive sensors, along with the use of a missile that is designed to intercept targets beyond the line-of-sight horizon of the firing platform. The committee was not briefed on any systems other than the Army's JLENS for solving this ASCMD problem.

With respect to OCMD, the committee observes that there is still no program that will provide a means for the ship-based defense of forces ashore against cruise missile attacks. Although ship-launched interceptor missiles of suitable range are available, the sensors that would permit them to engage cruise missiles not observable from the ship have not been developed or otherwise acquired. The Navy will have to develop the necessary airborne sensors to support an OCMD capability or seek ways in which systems of the other Services, such as JLENS, might be brought into position and employed.

**Recommendation: The Secretary of the Navy, the CNO, and the CMC should support the development of a competent cruise missile defense against anti-ship and overland cruise missiles. Beyond supporting the programmed development and acquisition of multifunction radar (MFR) and volume search radar (VSR), such a capability should include the following components:**

- **An elevated AMTI radar—possibly AWACS or unmanned aerial vehicle (UAV)-based—with robust overland clutter negation capabilities and with future capabilities to operate in a multistatic mode so that low-radar-cross-section overland targets can be engaged;**
- **An overland, over-the-horizon variant of the SM-2 missile with dual-mode, semiactive, and active terminal guidance; and**
- **The extension of cooperative engagement capability (CEC) to allow the employment of air-directed surface-to-air missiles (ADSAMs) against targets that are beyond the line-of-sight horizon of weapon launch platforms.**

The NAD program is designed to enable the defense of nearby forces against attack by shorter-range ballistic missiles. The committee observes that the program appears to be sound and adequately funded and that the necessary underlying R&D work is in place, including several improvements to SPY-1 signal processing that are also necessary for NTW. No significant deficiencies were noted.

The NTW program is not fully defined, and the portion defined so far is not completely funded. Although needed by the Navy for the defense of larger areas in a theater, including forces ashore, there is no funded program beyond the Aegis LEAP intercept effort funded mostly by congressional budgetary adjustments beyond those requested by the Navy. This effort includes the addition of a high-resolution-range unit to the SPY-1 radar, in addition to the upgrades made as part of the NAD program. In an effort to cope with the low funding priority that has been assigned to NTW by BMDO, Navy staff have laid out a spiral development concept, which is being implemented at a slow rate with small amounts of Navy and BMDO money. This concept, if fully funded, would provide for an interceptor missile and shipboard system upgrades that evolve from a basic ("contingency") NTW capability (Block IA). The deployable version to follow would not permit simultaneous conduct of other Aegis missions (Block IB). A final Block I capability (Block IC) would restore the ability to conduct all Aegis missions simultaneously.

An evolving NTW capability, designated Block II but not yet fully defined, would cope with the threat as it evolves in the next 10 to 15 years. This will require improvements to interceptor and radar performance, neither of which appears to be fully defined or funded as of yet. It appears possible that some system components will be developed incident to the cooperative program with Japan.

**Recommendation: Beyond supporting the SPY-1 upgrades to improve NAD and NTW discrimination capabilities, the Secretary of the Navy, the CNO, and the CMC should pursue an aggressive R&D effort aimed at producing the following capabilities:**

- **A high-resolution, X-band adjunct to the S-band SPY-1 radar that will allow discrimination among warheads, decoys, and debris and reduce the need for salvo launches;**
- **A hit-to-kill (HTK) vehicle with greater agility, divert capability, and lethal radius than the Block I HTK vehicle, giving it the ability to handle tethered and tumbling target complexes;**
- **A multicolor infrared sensor with improved sensitivity to extend acquisition ranges against low-infrared-signature targets and aid in discrimination; and**
- **A radar and/or LADAR on the hit-to-kill vehicle that could precisely measure body dynamics for effective discrimination against replica decoys.**

**Recommendation: In an effort to examine countermeasures beyond the design threat of naval theater ballistic missile defense systems, the Department of the Navy should maintain an ongoing red-blue effort that provides continuous analysis, design, and testing of potential theater ballistic missile**

**defense countermeasures and defense responses and works closely with corresponding Ballistic Missile Defense Organization (BMDO) efforts. This effort could be conducted in a manner similar to the prior Advanced Ballistic Reentry System Program, which developed penetration aids for U.S. intercontinental ballistic missile systems, or an extension of the current project Hercules, supported by BMDO, that is looking at advanced discrimination techniques.**

### 5.2.5 Current and Projected Marine Corps OCMD Capabilities

Marine Corps plans for OMFTS and STOM depend on shipboard basing of assault elements and rapid transport of light forces to inland objectives. The Navy is expected to provide air support—close air support along with Marine Corps air; combat air patrol; ship-based fire support; and ship-based early warning of and defense against air and ballistic missile attack. The Marine Corps is also dependent on the Navy for logistical support of many kinds. In the future, the Corps will have a ground-launched advanced medium-range air-to-air missile (AMRAAM) capability—a complementary low altitude weapon system—light enough to be taken ashore with assault units but with limited sensor capability, necessitating CEC cueing.

The current Marine Corps air defense radar, AN/TPS-59 (V3), has a large footprint and is not carried on board amphibious assault ships, and yet this is the only GBR that will be available to support Marines ashore or Navy ships off a coastline against the cruise missile threat until a new radar is developed late in this decade. The committee learned that a smaller mobile radar—the MRRS—is under consideration for future development and acquisition. Until this occurs, the committee observes that the STOM concept will be entirely dependent on robust shipboard sensors and missile defense capabilities with sufficient range to cover assault objective areas and weaken the threat to levels tolerable by the forces ashore.

**Recommendation: Recognizing that there will always be some gaps in naval air defense coverage due to extended littoral operations, the Secretary of the Navy, the CNO, and the CMC should support the development and acquisition of the complementary low altitude weapon system (CLAWS) and the multirole radar system (MRRS); interfaces should be developed for targeting and fire control to the following sensors:**

- **Army JLENS radar system,**
- **Marine Corps TPS-59 (V-3) radar system,**
- **E-2C RMP AMTI radar, and**
- **Air Force AWACS SPY-1/2 radar system.**

**Recommendation: Recognizing that the MRRS may not be ready in time to provide an initial targeting and fire control radar for the CLAWS, the Secretary of the Navy, the CNO, and the CMC should consider deployment of the TPS-59 radar on designated maritime preposition force squadrons as an interim measure.**

### 5.2.6 Battle Management Command, Control, and Communications (BMC3)

A commander must have the means to understand the operational environment, the location and condition of his forces, and the actions of the enemy. He/she must be able to communicate well enough to reallocate resources and vary subordinate assignments as appropriate to achieve his/her mission, keeping superiors advised as necessary. The need today to comprehend and control on a theater-wide basis presents an immense challenge.

The committee received several BMDO and Navy briefings about work on this subject. Although the various efforts seem necessary, clearly, they are not sufficient. The BMDO programs in this area do not appear to address the needs of a theater-level commander, and the Navy programs appear mostly to be concerned with the command and control of defensive systems. No concepts were presented for assembling an integrated picture of theater-level activity and presenting it to a commander in a useful way. As was noted above, no work seems to be under way to enable deconfliction of offense and defense operations. The committee observes that program effort in this area seems not to be coupled to thinking about a more open, network-centric communications architecture that would enable better access to information by lower-level participants, more useful reporting by those participants, and better decisions by the theater commander.

**Recommendation: Given that management of battle-space force components is a critical aspect of missile defense that is currently seriously deficient, Department of the Navy leadership should actively support efforts relating to doctrine, acquisition programs, and research to overcome such deficiencies, in particular by:**

- **Supporting current efforts such as the Single Integrated Air Picture (SIAP) System Engineering Office Program, which is seeking to enhance the quality of the air-space picture;**
- **Supporting the development of concepts of operations necessary for expeditionary and joint Service littoral operations, including means for offense-defense coordination;**
- **Recognizing that for success in these operations the Department of the Navy will require support from other Services; and**
- **Recognizing that all battle-space management development efforts must seek to accommodate the inclusion of unplanned force components.**

**Recommendation: Given that Link 16 and CEC, even when evolved and improved, will not provide a full battle management command, control, and communications (BMC3) capability for either overland cruise missile defense or theater missile defense, the Department of the Navy leadership should initiate actions leading to the development of a next-generation BMC3 system. This entirely new system, leveraging both commercial and defense technology advances, should include the following features:**

- **Support of highly flexible and adaptable combinations of naval and joint force configurations by allowing assets to interface readily with one another (e.g., through an Internet Protocol-based, quality-of-service-guaranteed infrastructure);**
- **Wide-bandwidth, bandwidth-on-demand wireless communication networks with dynamic allocation of resources; and**
- **Initial development of a prototype in parallel with existing BMC3 systems to encourage experimentation and adoption.**

**In addition, development of a high-bandwidth test bed would be particularly valuable. It would allow new capabilities to be tested and explored in the near term while the existing BMC3 systems continue to undergo their intended evolution; transition to the new capabilities would occur only after they had been adequately developed and accepted.**

### 5.2.7 Technology Investment

As presented to the committee by the Navy and Marine Corps, the developmental paths intended to evolve TMD capabilities are generally reasonable, although several exceptions are identified in this report. The evolutionary, or "spiral," development of added capabilities to pace the threat is a reasonable concept. However, the committee is concerned that the technology required to support the intended evolution is not being developed. The necessary investments must be made to bring the required technology to a state where it is available for use in the time frame intended.

**Recommendation: In its technology investment program, the Department of the Navy should develop sensors, weapons, and BMC3 architectures and algorithms that are adaptive and flexible enough to allow responding to unexpected threat capabilities and characteristics. These ballistic missile defense system elements should be combined into experimental systems for evaluation and refinement. The mature technologies from the program should be incorporated into future spirals of the NAD and NTW ballistic missile defense systems.**

# Appendixes

# A

# Acronyms and Abbreviations

| | |
|---|---|
| AADC | area air defense commander (module) |
| AADS70 | air defense systems of the 1970s |
| AAW | antiair warfare |
| ABL | airborne laser |
| ABM | antiballistic missile |
| ACA | airspace control authority |
| ACDS | advanced combat direction system |
| ACE | air combat element |
| ACM | attitude control motor |
| ACO | airspace control order |
| ADC | analog-to-digital converter |
| ADSAM | air-directed surface-to-air missile |
| AESA | airborne electronically scanned array |
| AIEWS | advanced integrated electronic warfare system |
| AIT | atmospheric interceptor technology |
| AMD | air and missile defense |
| AMRAAM | advanced medium-range air-to-air missile |
| AMRFS | advanced multifunction radio frequency system |
| AMTI | air moving target indication |
| ANTS | autonomous negotiating teams |
| ASCIET | all-Service combat identification evaluation test |
| ASCM | antiship cruise missile |
| ASCMD | antiship cruise missile defense |
| ASIC | application-specific integrated circuit |

| | |
|---|---|
| ATDLS | advanced tactical data link system |
| ATR | automatic target recognition |
| AWACS | airborne warning and control system |
| | |
| BAT | BMC4I advanced technology |
| BMC3 | battle management command, control, and communications |
| BMC4I | battle management command, control, communications, computing, and intelligence |
| BMD | ballistic missile defense |
| BMDO | Ballistic Missile Defense Organization |
| | |
| C2P | command and control processor |
| C4ISR | command, control, communications, computing, intelligence, surveillance, and reconnaissance |
| C&D | command and decision |
| CAP | combat air patrol |
| CC&D | common command and decision system |
| CCID | composite combat identification |
| CEC | cooperative engagement capability |
| CEP | cooperative engagement processor |
| CID | combat identification |
| CINC | commander in chief |
| CIWS | close-in weapon system |
| CLAWS | complementary low altitude weapon system |
| CMC | Commandant of the Marine Corps |
| CMD | cruise missile defense |
| CNO | Chief of Naval Operations |
| COABS | control of agent-based systems |
| COIL | chemical oxygen-iodine laser |
| CONOPS | concept of operations |
| COTS | commercial off-the-shelf |
| CRD | capstone requirements document |
| | |
| DACS | divert and attitude control system |
| DARPA | Defense Advanced Research Projects Agency |
| DASN | Deputy Assistant Secretary of the Navy |
| DDS | data distribution system |
| Dem/Val | demonstration/validation |
| DNMS | dynamic network management system |
| DOD | Department of Defense |
| DOE | Department of Energy |
| DSP | Defense Satellite Program |

| | |
|---|---|
| ECCM | electronic counter-countermeasure |
| ECM | electronic countermeasure |
| EIT | exo-atmospheric interceptor technology |
| EMD | engineering and manufacturing development |
| EOC | early operational capability |
| ERGM | extended-range guided missile |
| ERINT | extended-range interceptor |
| ERIS | exo-atmospheric reentry vehicle interceptor system |
| ESM | electronic support measure |
| ESSM | evolved sea sparrow missile |
| EV | experimental version |
| EW | electronic warfare |
| | |
| FABMDS | field army ballistic missile defense system |
| FEL | free electron laser |
| FEWS | follow-on early warning system |
| FIOP | Family of Interoperable Operational Pictures program |
| FLIR | forward-looking infrared |
| FNC | future naval capability |
| FOS | family of systems |
| FPA | focal plane array |
| FRP | full-rate production |
| FUE | first unit equipped |
| | |
| GBR | ground-based radar |
| GCCS-M | global command and control system-maritime |
| GLONASS | Russian equivalent of GPS |
| GMTI | ground moving target indication |
| GPS | global positioning system |
| | |
| HEL | high-energy laser |
| HF/DF | hydrogen fluoride/deuterium fluoride |
| HMMWV | high-mobility multipurpose wheeled vehicle |
| HTK | hit-to-kill |
| | |
| ICBM | intercontinental ballistic missile |
| IFOG | interferometric fiber optics gyro |
| IMU | inertial measurement unit |
| IOC | initial operational capability |
| IP | Internet Protocol |
| IR | infrared |
| IRBM | intermediate-range ballistic missile |
| ISO | Information Systems Office (DARPA) |

| | |
|---|---|
| ISR | intelligence, surveillance, and reconnaissance |
| ITO | Information Technology Office (DARPA) |
| | |
| JDN | joint data network |
| JFACC | joint force air component commander |
| JFC | joint force commander |
| JICO | joint interface control officer |
| JLENS | joint land attack cruise missile defense elevated netted sensors |
| JROC | Joint Requirements Oversight Council |
| JSTARS | joint surveillance and target attack radar system |
| JTAMDO | Joint Theater Air and Missile Defense Organization |
| JTIDS | joint tactical information distribution system |
| JVMF | joint variable message format |
| | |
| KEW | kinetic-energy weapon |
| KKV | kinetic-kill vehicle |
| KV | kill vehicle |
| | |
| LADAR | laser detection and ranging (laser radar) |
| LAN | local area network |
| LIDAR | laser identification and ranging |
| LMT2 | Link 16 missile and tactical terminal |
| LWIR | long-wavelength infrared |
| | |
| MDAP | major defense acquisition program |
| MEADS | medium extended air defense system |
| MEB | Marine Corps expeditionary brigade |
| MEMS | microelectromechanical system |
| MFR | multifunction radar |
| MIDS | multifunctional information distribution system |
| MMIC | microwave/millimeter-wave monolithic integrated circuit |
| MMW | millimeter-wave |
| MPF | maritime preposition force |
| MRRS | multirole radar system |
| MSI | multisensor integration |
| MWIR | mid-wavelength infrared |
| | |
| NAD | Navy area defense (system) |
| NAVSEA | Naval Sea Systems Command |
| NMD | national missile defense |
| NSSMS | NATO sea sparrow missile system |
| NTMD | Navy theater missile defense |
| NTW | Navy theater wide (system) |

| | |
|---|---|
| OCM | overland cruise missile |
| OCMD | overland cruise missile defense |
| OMFTS | Operational Maneuver From the Sea |
| ONR | Office of Naval Research |
| OPEVAL | operational evaluation |
| OPNAV | Office of the Chief of Naval Operations |
| ORD | operational requirements document |
| | |
| P3I | preplanned product improvement |
| PAC-3 | Patriot advanced capability-3 |
| PEO TSC | Program Executive Office for Theater Surface Combatants |
| PPLI | precision position location information |
| | |
| QOS | quality of service |
| | |
| R&D | research and development |
| RAM | rolling airframe missile |
| RCS | radar cross section |
| RF | radio frequency |
| RMP | Radar Modernization Program |
| ROE | rule of engagement |
| RV | reentry vehicle |
| | |
| SA | selective availability |
| SAR | synthetic aperture radar |
| SBEV | space-based experimental version |
| SBIRS | space-based infrared sensor |
| SDI | strategic defense initiative |
| SDIO | Strategic Defense Initiative Organization |
| SENSIT | sensor information technology |
| SIAP | single integrated air picture |
| SIGINT | signal intelligence |
| SLS | shoot-look-shoot |
| SM | standard missile |
| SPAWAR | Space and Naval Warfare Systems Command |
| SRBM | short-range ballistic missile |
| SSDS | ship self-defense system |
| STAP | space-time adaptive processing |
| STOM | Ship-to-Objective Maneuver |
| | |
| TAD | theater air defense |
| TADIL | tactical digital information link |
| TAMD | theater air and missile defense |

| | |
|---|---|
| TBM | theater ballistic missile |
| TBMD | theater ballistic missile defense |
| TCN | tactical component network |
| TCS | theater combat system |
| TCT | theater collaborative tracking |
| TDMA | time division multiple access |
| TEWA | threat evaluation and weapon assignment |
| THAAD | theater high altitude area defense |
| THEL | tactical high-energy laser |
| TMD | theater missile defense |
| TOM | target object map |
| TPS | transportable (pulse) radar surveillance system |
| T/R | transmit/receive |
| TRAP | tactical receive applications program |
| TRE | tactical receive equipment |
| TSB | time slot block |
| TSC | theater surface combatant |
| TVM | target-via-the-missile |
| | |
| UAV | unmanned aerial vehicle |
| UEWR | upgraded early warning radar |
| | |
| $V_{bo}$ | velocity at burnout |
| VLS | vertical launch system |
| VSR | volume search radar |

# B

# Analysis of the Capabilities and Limitations of Link 16

The Navy is currently supporting programs that are dedicated to improving the performance of Link 16 tactical radio network technology. These programs are managed by the ATDLS program office (SPAWAR PMW 159). The committee finds that these programs have technical merit and are likely to be of substantial benefit to the Navy. However, they are best viewed as late-life upgrades to a system that is nearing the end of its technical life cycle. This appendix contains the detailed technical analysis of Link 16 that has led the committee to its conclusions. Table B.1 shows the characteristics of networking schemes currently used in the Navy's tactical arena.

JTIDS and its slightly more modern variant, the multifunctional information distribution system (MIDS), are the Navy's chosen radio subsystems for distributing force control messages. As such, these radio subsystems implement the Link 16 (TADIL-J) networking scheme and message set. The messages include surveillance tracks, weapons coordination, air control, target information, PPLI, and even digitized voice networks. JTIDS radios—or their MIDS variants—will be installed on a variety of aircraft, surface ships, and submarines over the next 7 years, as well as in Patriot and THAAD forces.

## B.1 JTIDS CHARACTERISTICS

Certain technical characteristics of the JTIDS waveform that have important effects on the types of networks that can be built with JTIDS radios are briefly described in the following paragraphs.[1]

[1]The material in this section is based on Naval Studies Board, National Research Council. 2000. *Network-Centric Naval Forces: A Transition Strategy for Enhancing Operational Capabilities*, Na-

TABLE B.1 Characteristics of Networking Schemes

| Characteristic | TADIL A Link 11 | TADIL C Link 4A | TADIL J Link 16 | TADIL J Link 22 |
|---|---|---|---|---|
| Antijam | No | No | Yes | No |
| Crypto-secure | Yes | No | Yes | Yes |
| Data rate (kbps) | 1.3 to 2.25 | 5.0 | 28.8 to 115.2 | 2.4 |
| Message standard | M series | V/R series | J series | J series |
| Participants | 20 | 4-8 | 128+ | 40 |
| Critical nodes | Yes | Yes | No | No |
| Voice circuits | No | No | 2 | No |
| Architecture | Radio broadcast | Radio point-to-point | TDMA | TDMA |
| Frequency | HF/UHF | UHF | UHF/Spread | HF/UHF Spread |

SOURCE: Welch, LCDR David, USN, "TADIL Comparison" in "U.S. Naval Tactical Data Links," briefing to the Tactical Network Panel of the Committee on Network-Centric Naval Forces on February 17, 1999, Command and Control Systems Directorate, Office of the Chief of Naval Operations (N62G), Washington, D.C.

### B.1.1 Waveform

JTIDS operates in the L-band. It divides the spectrum into 51 channels between 969 MHz and 1209 MHz, with a channel spacing of 3 MHz. Certain portions of the spectrum are also used for identification, friend or foe (IFF), tactical air and navigation (TACAN), distance measuring equipment (DME), and Mode S, which excludes two subbands and imposes some restrictions on exactly how JTIDS can be used in noncombat situations. In particular, time-slot duty cycles for JTIDS must be restricted to no more than 20 percent under normal conditions. Exercise conditions do not have duty-cycle restrictions, and full combat conditions have no restrictions.

JTIDS uses a TDMA waveform. Every 24-hour day is divided, in the JTIDS waveform, in 112.5 epochs. Each epoch lasts 12.8 min and is subdivided into 64 frames of 12 s apiece. Each frame is further subdivided into 1536 time slots.

tional Academy Press, Washington, D.C. Information on JTIDS has been derived from two sources: (1) Welch, LCDR David, USN, "U.S. Navy Tactical Data Links," briefing to the Tactical Network Panel of the Committee on Network-Centric Naval Forces on February 17, 1999, Command and Control Systems Directorate, Office of the Chief of Naval Operations (N62G), Washington, D.C., and (2) U.S. Army Program Executive Office Air and Missile Defense and Life Cycle Engineering Center, Missile Research Development and Engineering Center, U.S. Army Aviation and Missile Command, "Introduction to JTIDS," Redstone Arsenal, Ala.

Each time slot is thus 7.8125 ms long. Time slots within frames are organized into three distinct sets labeled A, B, and C. Time slots within a frame are identified as A-0, B-0, C-0, A-1, B-1, C-1, . . . A-511, B-511, C-511. A given radio ("terminal") may have up to 64 blocks of time slots assigned to it. Each time-slot block is defined by a triplet called a time slot block (TSB): set (A, B, or C); index (0 to 511); recurrence rate (0 to 15). Each assignment for a given terminal is designated as transmit, receive, or relay.

A JTIDS net is a group of terminals that exchange messages among themselves. In other words, it is a group of terminals whose time slots have been defined so that when one member of a net is transmitting, every other member of the net is receiving. Obviously this requires careful planning to ensure that indeed all the other members are receiving at that time, that only a single radio is granted a transmit time slot at a given time, and so forth. The JTIDS architecture allows 127 different nets (numbered 0 through 126) to be active simultaneously within the same RF spectrum. Since JTIDS is a frequency-hopping radio, each net is made mutually exclusive by assigning a unique frequency-hopping pattern for transmissions.

### B.1.2 Other Technical Characteristics

- *Access modes.* As defined, JTIDS provides three distinct access modes for a terminal that needs to transmit: dedicated access, contention access, and time slot reallocation access.
- *Dedicated access.* In this mode, the network planners ensure—by preparing the corresponding time-slot plan for a given network—that a given JTIDS terminal has exclusive use of an assigned TSB. This mode has the advantage that the terminal is guaranteed access to the network at regular intervals; it also has the corresponding disadvantage that the time slot is wasted if the terminal has nothing to say at a given moment.
- *Contention access.* In this mode, which is quite different, a given net provides a pool of time slots for any terminal's use. Any terminal that needs to transmit will randomly select a time slot from this pool and transmit in that time slot. This mode has a number of advantages: it is easy to plan, makes it quite simple for terminals to enter or leave the net while the net is in operation, and provides some of the traffic efficiencies of statistical multiplexing for traffic that is bursty or hard to predict. Its main disadvantage is that multiple terminals may transmit during the same time slot, which can result in lost messages and/or some terminals hearing one transmitter while others hear a different one.
- *Time slot reallocation access.* In this, the most complex mode, all terminals share a single pool of time slots, as is also true for contention access. Rather than transmit at will, however, the terminals perform a distributed algorithm to apportion the time slots. Each terminal transmits its bandwidth needs periodical-

ly, and every terminal performs identical algorithms to ensure that the pooled time slots are apportioned according to the needs. The committee believes that this access scheme has not yet been implemented in practice, but—as will be seen below—it is one of PMW 159's projects.

### B.1.3 JTIDS Data Rates

Each JTIDS time slot has the following components: The time slot begins with a variable-start jitter delay; then, synchronization and time-refinement patterns; the payload (message header and data); and, finally, dead time to allow for RF propagation. This discussion will concentrate on the message data portion of a time slot. Each data portion can contain 3, 6, or 12 75-bit words, depending on the exact encoding of the message. Thus, each time slot can carry anywhere from 225 to 900 bits of data payload, giving an aggregate data rate for a given JTIDS net of between 28,800 and 115,200 bps. Some of this raw capacity is used for housekeeping and so is not available for tactical traffic, but these numbers give an idea of the approximate capacity of a JTIDS net.

By comparison, current commercial phone-line modems run at roughly 53,000 bits per second in the downstream direction. Thus, one JTIDS net has a raw capacity ranging from one half to twice the capacity of a phone-line modem. Since JTIDS divides its available L-band spectrum into 51 channels, the extreme upper bound on the number of bits per second that can be transmitted simultaneously from all JTIDS terminals in a tactical arena is 51 × 115,200, or 5,875,200 bps. This assumes that all available spectrum is devoted to JTIDS, that all terminals use the maximum possible data rate, and that all time slots in all channels are used for transmission, and it ignores the overhead of housekeeping bits. Working from the previous calculation, JTIDS achieves 5,875,200 bps in 51 × 3 MHz of RF spectrum, for an aggregate spectral efficiency of 0.0384 bps/Hz. Partly, of course, this is driven by the tactical need for very robust antijam features. To a noticeable extent, though, it is driven by the basic short-frame TDMA structure of the JTIDS waveform, where rather short payloads are surrounded by the dead times of synchronization patterns and propagation allowances.

## B.2 ASSESSMENT OF PMW 159'S PLANNED IMPROVEMENTS TO LINK 16

Table B.2 presents the improvements in tabular overview form. Subsequent paragraphs describe and assess each of the improvements in more detail.[2]

[2]Information in this section is derived from McCloud, Kenneth L., "PMW 159 Advanced Tactical Data Link Systems (ATDLS) Program Office," briefing to the committee on July 26, 2000, Space and Naval Warfare Systems Command (PMW 159A), Arlington, Va.

TABLE B.2 Planned Improvements and Potential Benefits

| Planned Improvement | Potential Benefit |
|---|---|
| Dynamic network management system (DNMS) for Link 16 | Incremental increases in the flexibility of Link 16 networks, perhaps coupled with greater ease of planning and configuring for such networks |
| Enhanced throughput | Higher bandwidth communications across Link 16 radio channels |
| Optimized relative navigation | More accurate relative position and time information for Link 16 platforms |
| Joint range extension, S-TADIL J | Increased ability to transmit J-series messages across non-JTIDS radio channels |
| Link 16/JVMF advanced concept technology demonstration | Gateways between Link 16 radios and their messages, on the one hand, and the Army's messaging system on the other |
| Link 16 missile and tactical terminal (LMT2)/TacLink weapons | Tactical command and position/location links to guided munitions |

### B.2.1 Dynamic Network Management System for Link 16

These are a set of interrelated changes to the JTIDS channel access protocols that should allow more flexible use of JTIDS networks. Key technical features are improvements in the methods used for late net entry, for reallocating time slots as demand changes, and for varying the throughput rate dynamically. In passing, it is noted that many of these techniques have already been used in a wide variety of other radio systems and hence pose relatively little technical risk.

*Assessment.* These incremental improvements will probably succeed and will make JTIDS somewhat more flexible than with its current (highly rigid) architecture. As a result, there is potential to make JTIDS easier to use in practice. This is important and useful work and should be supported. On the other hand, to a large extent these improvements are merely Band-Aids for a fundamentally unsuitable network architecture. They will allow more flexibility in the use of JTIDS networks, but the improved system can be considered "flexible" only in comparison with classic JTIDS; it is by no means as flexible as modern commercial systems. In addition, of course, JTIDS will remain a closed system. The final judgment, therefore, is that this work should be supported—it is certainly better than classic JTIDS—but that it will not in the end provide the degree of flexibility required for today's or tomorrow's tactical communication needs.

### B.2.2 Enhanced Throughput

This program aims to increase JTIDS's bandwidth by employing modern channel encoding techniques to achieve more bits per second per hertz. The upper bound on the improved speed is claimed to be 1.1 Mbps, which is nearly 10 times the current maximum rate.

*Assessment.* This program is a low-risk incremental improvement to JTIDS that may well have practical utility. As such, it should be supported. It would be unwise, however, to assume that the new maximal rate of 1.1 Mbps will in fact be achieved often in practice. Maximum rates for wireless communications are usually achieved only for stationary objects that are quite close to each other in a clear RF environment, because performance degrades quickly with Doppler effects, distance, and interference. Since JTIDS is generally employed between mobile platforms across relatively long distances, the actual data rates may be well below maximal. In addition, since most JTIDS time slots are received by a number of different platforms, the transmitted data rate must reflect the lowest common denominator among the receivers (e.g., the farthest away, the fastest moving, the one with the oldest equipment). Again, the reader is reminded that the commercial wireless world is in a creative foment at the moment and that a large number of very-high-speed wireless technologies are now appearing in the market. On the whole, these technologies are likely to deliver significantly higher overall throughput than enhanced JTIDS since they do not suffer from JTIDS' very short time slots, which ensure that a very high percentage of potential transmission time is in fact sacrificed to dead time between bursts. This is, therefore, a good incremental enhancement to JTIDS, but the Navy should also look elsewhere for high-bandwidth wireless technology.

### B.2.3 Optimized Relative Navigation

This program plans to improve the relative navigation capabilities of Link 16 so that it will deliver position/location information with an accuracy that equals or exceeds that of the current GPS system ($\leq$3 m circular error probability) and time synchronization to the nanosecond level. It will do so by transmitting raw, uncorrected pseudo ranges and employing new algorithms on these data.

*Assessment.* Higher-level functions of battle management such as the SIAP rely on highly accurate position and time information. Thus, any effort to improve these data could have a significant payoff. This particular method has the additional virtue of being independent of GPS and thus providing a robust backup capability for position and time services and should be supported.

### B.2.4 Joint Range Extension, S-TADIL J

Joint Range Extension, S-Tadil J combines two distinct programs, both of which aim to add a capability to transmit J-series messages across non-Link-16 communications channels.

*Assessment.* This is a highly desirable goal, but the approach is fundamentally misguided. The key problem here is that JTIDS has confounded the distinct problems of message formats and RF channel architecture. The proper solution—and one that has been universally adopted in the commercial communications world, both in the Internet and in all telephone technology, for decades—has been to use a layered protocol stack so that any type of message can flow across any type of communications medium. Rather than sort out how J-series messages should be conveyed across any type of medium, these programs are attempting minor incremental "kludges" to work across satellites and so on. The committee believes that this entire approach will ultimately reach a dead end. A program that determined how J-series messages could sent using the Internet Protocol suite would better serve the Navy.[3] They could then be transmitted across virtually every known type of communications channel with no additional effort on the Navy's part.

### B.2.5 Link 16/Joint Variable Message Format ACTD

This advanced concept technology demonstration (ACTD) will show that the J-series messages conveyed across Link 16 can be translated to the messages formats employed in the Army's digitized battlefield.

*Assessment.* From a high-level viewpoint, it is unfortunate that the Army message formats are not compatible with those used in Link 16, but since this is the case it is clearly better to gateway the two systems together with translators than to have no connection between them at all.

### B.2.6 Link 16 Missile and Tactical Terminal/TacLink Weapons

This concept envisions a small tactical radio, based on Link 16 technology, that can be installed in cruise missiles and other guided munitions to give them (1) a precise positioning system, (2) a command link for updates on mobile target locations, and (3) improved potential for battle damage assessment. This concept would also enhance overall situational awareness since it would allow all missiles and guided munitions to be included in the SIAP.

*Assessment.* Judgment is reserved on this concept. It seems to be a forward-thinking idea that is well aligned with the necessary future direction of

[3]There have been some limited experiments along these lines.

BMC3. And JTIDS certainly does have good antijam properties, which would be essential in such tasks. On the other hand, JTIDS provides a poor starting point for this concept, partly because JTIDS networks have proven extremely difficult to plan and configure, but mainly because thus far JTIDS radios are extremely expensive. It is certainly possible that both problems could be overcome, but the solution would certainly be much easier if a different starting point were adopted.

# C

# Obtaining More Flexible BMC3 Configurations

The main body of this report makes the important point that greater flexibility in establishing missile defense BMC3 configurations is necessary and that commercial wireless communications technology provides a critical ingredient in obtaining this flexibility. This appendix elaborates on that point by first discussing a system engineering process that would lead to greater flexibility and then providing some detail on commercial wireless communications technology.

## C.1 SYSTEMS ENGINEERING FOR BMC3

### C.1.1 Introduction

The committee is concerned that the BMC3 structures envisioned, as presented in the various briefings it received, are too rigid and do not recognize the need for flexibility and adaptability that is necessitated by the ad hoc nature of likely deployments to hostile areas. The committee is also concerned that concepts of operations were not presented and seem not to have been developed for BMC3 functions.

The committee heard repeatedly that, for most major conflicts and exercises in the last decade, the BMC3 system was jury-rigged. While this may appear the exception, it is likely to be the norm. The reason is that in the Navy (as well as the other Services) the operator and the acquisition components live in different worlds.

The operator participating in a conflict is in a fluid situation. He uses whatever resources are available. Often he does not have some resource that he might have expected and will occasionally be presented with resources or capabilities that were not expected. Those unexpected resources invariably were not anticipated in the systems engineering process, from either a concept of operations viewpoint or a technical interface viewpoint. To the extent time and technical capabilities are available, the operator will find a way to accommodate these unplanned resources. This situation is even more amplified in joint operations. When the Joint CINC gathers his forces, he will attempt to piece together capabilities from the available resources. Often he will perceive an opportunity to connect complementary capabilities that were not products of common systems engineering.

Acquisition people live in the world where systems engineering is constrained to encompass only those capabilities that are based on validated requirements and priorities. These requirements are usually based on a concept of operation that makes assumptions about the resources and threats that might be justified 5 to 10 years in the future. These people are further constrained by the myriad of legacy systems with which they must interoperate. There is ample justification for such a process in a world of predictable threats, understood needs, and constrained resources. However, this traditional systems engineering approach will not yield a BMC3 system that meets the Navy's needs for future theater missile defense because of uncertainties in the projections of threats and of the capabilities and technologies needed for managing TMD. The time cycles for evolution of the technology and for evolution of the threats are both inside the systems engineering cycle. Furthermore, as was previously described, for the Navy to participate effectively in theatre missile defense, it must accommodate systems outside its control.

This leads to the conclusion that the Navy must plan for the system to be jury-rigged. More precisely, the Navy needs a BMC3 architecture that accommodates the introduction of unplanned resources and capabilities. This does not imply abandoning systems engineering or abandoning the requirements process. Rather it implies development of an open BMC3 systems architecture within which the systems engineering and priority setting must take place—an architecture that plans for unanticipated capabilities that will be introduced at a later time.

The difference between jury-rigging and having an architecture into which new capabilities may be introduced is profound. While both require creativity, the difference is how the creativity is used. In jury-rigging, creativity is wasted in figuring out how to pass information and in solving timing differences between two systems that were designed with different architectures. In the open systems approach, the creativity is used to choose the protocols, standards, and interface definitions applied to the process of building value-added capabilities that facilitate the identification, correlation, and other BMC3 functions.

### C.1.2 The Importance of Architecture

Much has been said about system architecture in recent years. The Defense Science Board and the individual Service advisory boards have dealt with the subject. DOD has even established offices with responsibility for defining specific architectural principles and processes, such as the Open Systems Joint Task Force.

In general, an architecture describes the components and their relationship to one another. These may be defined in terms of the operational architecture, the systems architecture, and the technical architecture. The Army Science Board 1994 Summer Study[1] elaborated on these three architectural views as follows:

- *Operational architecture.* A description, often graphical, of the required connectivity between force elements: operations facility to weapon systems, sensors to shooters, and so on. The description defines who will communicate with whom (voice and data) and includes the type, timeliness, and frequency of the information sent between these elements.
- *Systems architecture.* A description, including graphics, of the technical characteristics and the interconnection of all parts of an information system. The description identifies all system elements (radios, telecommunication switches, computers, and so on) and specifies the bandwidth required between each element, the electrical interfaces on each element, schematics for hardware, software specifications, and so on.
- *Technical architecture.* A minimal set of rules (e.g., protocols, standards, software interface specifications) governing the arrangement, interaction, and the interdependence of the parts or elements that together may be used to form an information system. Its purpose is to ensure that a conformant system satisfies a specified set of requirements (e.g., interoperability, portability, and survivability).

In each case, it is critical that the interfaces be well defined so that as components or concepts change, the amount of effort to accommodate those changes is contained. These interfaces are often bound into standards once the interfaces are sufficiently broad and tested.

### C.1.3 Internet Technology

Fortunately, the Internet provides an example of an open system in which the protocols and interface standards have already been worked out. Not only is the technology understood and tested, but robust products are also readily available from commercial sources at affordable prices. Finally, substantial commercial investments in products will make significant bandwidth achievable in a wireless environment.

---

[1]Army Science Board. 1995. *Technical Information Architecture for Command, Control, Communications, and Intelligence,* 1994 Summer Study, Office of the Assistant Secretary of the Army (Research, Development, and Acquisition), Washington, D.C., April.

It is not suggested that the Navy depend on the Internet for TMD BMC3, although it might in some future engagement exploit the Internet if it is available. It is suggested that the Navy use Internet technology—the protocols, standards, and supporting commercial products embodied in the Internet—as the basis for its BMC3. Within this architecture, the Navy can use the traditional systems engineering and requirements processes to ensure that the systems that need to participate in BMC3 are able to and those that should not participate are blocked using some combination of security technology.

Clearly, Internet-based technology does not solve the BMC3 problem. It simply provides the high-bandwidth, open systems framework for value-added capabilities to be developed as nodes on the net. It provides the mechanisms for rapidly including new capabilities such as sensors, track correlation techniques, and discrimination methods to be added in a planned rather than jury-rigged fashion. It opens the door to a combination of push-and-pull techniques to be developed and tested through exercises.

Use of the Internet is not foreign to Navy thinking. Indeed the Navy has a number of Internet technology efforts under way. The committee simply suggests that the Navy leverage the results of these efforts in evolving its TMD BMC3. It is also noted that the Air Force is experimenting with the Internet through its battle-space infosphere efforts, including AWACS, which means that a future TMD BMC3 using Internet technology could readily incorporate AWACS sensor data.

### C.1.4 The Transition Path

These observations present the Navy with a dilemma. The Internet-technology-based approach is a radical departure from the legacy systems and planned improvements. The committee recognizes the enormity of the task if the Navy were to simply abandon its current systems and launch a new Internet-technology-based approach. On the other hand, the planned improvements to current systems are incremental in nature and will not position the Navy for the future TMD BMC3, which requires flexibility.

In essence, it is suggested that the Navy leapfrog the current technology, which is nearing the end of its life cycle, onto an infrastructure technology that is still at the beginning of its life cycle.

There is an affordable strategy that will allow the Navy to maintain legacy systems and their incremental improvements while migrating to a more open-system, Internet-technology-based solution for BMC3. It will require additional funding but in the long run will enable the Navy to achieve the kind of flexibility it needs in a much more timely and cost-effective manner. The strategy is outlined in the next section.

### C.1.5 Build a Succession of Prototype Internet-Technology-based Infrastructures

In parallel with the evolutionary improvements planned for legacy systems, begin prototyping an infrastructure based on commercially available technology. Recognize that the commercially available products used for the initial prototype will mature quickly and will need to be replaced several times before they are committed to the field. The successive prototypes can be experimented with during exercises. The Navy should fund the interfacing of legacy systems and require that all new systems interface to the prototypes so that during exercises the prototype implementation may be stressed. In this process, it is important not to let the infrastructure stray from the evolving Internet. Finally, development of BMC3 capabilities should be encouraged insofar as they add value to the prototype infrastructure and can solve problems experienced in exercises.

It is important that these prototypes represent a continuously upgraded series of capabilities. Maximum advantage should be derived from experimentation and encouraging research results to be added as nodes on the net, which can be evaluated by operational forces. Efforts should be made to include the other services in these experiments.

The committee fully understands that current products will not meet Navy requirements in areas such as antijam, security, and real-time performance that the Navy's legacy systems currently meet. However, the technology is moving so rapidly that the committee expects some of those requirements to be exceeded within 3 to 5 years. The Navy can invest in the military-unique requirements that will not be satisfied by commercial technology. This approach will allow the Navy to significantly reduce the lifetime of the legacy systems and avoid the predictably high cost of ownership.

### C.1.6 Continue to Evolve Legacy Systems Incrementally

As previously described, the JTIDS/Link 16 approach is a bandwidth limited, rapidly obsolescing technology that will impede future operational flexibility. There are a variety of planned improvements that may make it somewhat more effective, and these should be continued as planned. However, at each stage, the Navy should evaluate the utility and cost of the improvements against the evolving capability provided by the Internet technology prototyping. The goal should be to use JTIDS/Link 16 when nothing better is available but to wean the BMC3 system from depending on it.

CEC is an excellent implementation of the philosophical approach advocated by the committee in that it seeks to accommodate distributed sensors. It provides the basis for the current self-defense capabilities and gives the Navy some area defense capability. It is, however, a closed-loop system that will not provide the long-term capabilities needed for a more complete TMD BMC3.

The Navy should continue the approach without locking itself into the protocols and standards imposed.

### C.1.7 Maintain Parallel Paths Until the Transition Is Complete

This approach means pursuing a dual path for some time. By requiring that all new system capabilities interface to the prototype, the Navy will ensure its ability to transition gracefully to the Internet technology at the earliest possible point and avoid long-term legacy costs.

## C.2 WIRELESS CONNECTIVITY

### C.2.1 Current and Near-future Commercial Trends

The overall BMC3 problem is a tangle of many technical subproblems. These include agreement on the identity of objects in the battle space, the deconfliction of airspace on an as-needed basis in real time, the assignment of sensors in response to changing conditions, decisions concerning which interceptor or interceptors should aim for which incoming missiles, and so forth. These are difficult problems and historically they have been further compounded by the great scarcity of communications bandwidth between platforms in the battle space and the need for assured tactical data distribution over relatively poor radio channels between these platforms. Thus for many years tactical information systems have grappled with the extremely difficult "subject matter" problems of BMC3 within the additional, and very severe, constraints imposed by issues of tactical radios and their meager bandwidth.

The "subject matter" problems of BMC3 remain very difficult. However, the very rapid rise of new commercial technologies in wireless communications, and particularly in wireless Internet communications, brings a brand new opportunity to disentangle these "subject matter" problems from the rather distinct problems of radio connectivity and communications channels.

Put briefly, the wireless communications world is at present moving very quickly from an economics of scarcity to one of abundance. The Navy should move quickly to capitalize on this new opportunity, because it will allow the partitioning of the almost intractable BMC3 problem into two easier subproblems—information processing ("subject matter") and connectivity ("radios")—and will, for the near-term future, reduce the basic connectivity issue to one that admits a relatively straightforward solution. Thus the Navy will be able to concentrate more on the information processing aspects, which are the harder aspects, in an environment that is relatively unconstrained in its use of wireless communications bandwidth between distributed platforms. This is an enormous change from the situation just a few years ago. Detailed market estimates for wireless Internet access are not available to the committee, but Killen & Associates forecast a 71 percent compound annual growth rate for this market, from $1.3 billion in 1998 to $19.2 billion in 2002.

To put a less-speculative dollar figure on the commercial interest in wireless Internet communications, the most recent auction of radio-frequency (RF) spectrum rights in the United Kingdom brought in more than $30 billion and one in Germany brought in $45 billion.[2] That is, telecommunications service providers have recently spent a total of $75 billion to acquire rights to use certain regions of the RF spectrum within the United Kingdom and Germany. They will, of course, spend large additional sums on equipment and real-estate leases to build the infrastructure they need to provide wireless Internet connectivity. Bidding for spectrum rights in other countries is expected to be just as expensive.[3]

It is safe to say that there is enormous commercial interest in wireless Internet services and that it will be difficult for the Navy to match the investments that are currently being made in the commercial arena. Fortunately it does not have to; on the contrary, it can leverage them for its own uses.

Whether these commercial advances in wireless Internet technology have any relevance for the Navy and its tactical systems is considered next. As will be seen, they certainly are highly relevant and promise great utility for the Navy's tactical information systems. Perhaps the best way to approach the Navy's specific needs for wireless communications is to give a brief recap of its technical requirements, mapping each of the requirements onto the current commercial technology. When the wireless medium is thought of as a communications service that allows tactical platforms to communicate with one another, it is clear that four key technical issues must be addressed:

- Quality of service (QOS),
- Bandwidth,
- Flexibility, and
- Military-specific characteristics.

The remainder of this section describes each of these issues briefly and shows that the first two are extremely important in the commercial telecommunications industry: they are currently receiving very substantial investments and

---

[2]For auction information, see Broadband Fixed Wireless Access Spectrum Auction Site of the Radiocommunications Agency, United Kingdom, at <www.spectrumauctions.gov.uk/>, and (2) Xinhua News Agency, 2000, "Roundup: Mobile Commerce Emerging as New Business Trend," Special Editions, Northern Light Technology, Inc., Cambridge, Mass., September 9, available online at <http://special.northernlight.com/wireless/roundup.htm>.

[3]Indeed, these auctions raise issues in their own right for the Navy. As it happens, JTIDS radios currently occupy a highly desirable swath of RF spectrum. It is not beyond the bounds of possibility that the DOD would lose access to this spectrum if it were auctioned off to the highest commercial bidders.

indeed are already being deployed in a major way. The third issue, flexibility, is receiving attention in the commercial world but is by no means perfect. The fourth issue includes all the military-specific problems in wireless communications (antijam is an example) and so will require military investment, as has historically been the case. The really good news, however, is that the two most difficult problems—QOS and bandwidth—have been tackled with great vigor in the commercial world, and the Navy's tactical communications can be the beneficiary.

### C.2.2 Quality of Service

QOS is most readily understood in terms of specific services that must be provided with high degrees of reliability. In general, tactical uses for QOS demand high availability, low-loss and low-delay bounds, and often have military precedence or priority.

It is interesting to note that commercial demand for voice over Internet Protocol (IP) led in the past year to readily available technology for this capability. The extent of this revolution is perhaps not yet apparent outside the telecommunications industry, but it is indeed remarkable. Every major telecommunications company is deploying a voice over IP infrastructure as its next-generation telephony system. As has been widely reported in the press, AT&T has ceased buying conventional circuit switches. AT&T's chairman, Michael Armstrong, has expressed the company's telephony plans very succinctly: "For AT&T, it's IP."[4] Equipment vendors are similarly committed to voice over IP. The list of such vendors includes all major manufacturers of telephony equipment (Lucent, Ericsson, Nokia, Motorola, Nortel, and so on), all major manufacturers of computer and data networking equipment (Cisco, Microsoft, IBM, Compaq, Sun, 3Com, and so on), and all major component manufacturers (Intel, Texas Instruments, and so on). All these companies have QOS-enabled Internet products currently available for sale.

Frost & Sullivan's estimates show voice over IP telephony services bringing in about $1 billion in 2000 and rising to more than $90 billion by 2006. Voice over IP equipment sales are expected to accelerate at a similar rate. Although it may not be immediately apparent to anyone outside the telecommunications industry, the near-term future of QOS networks is now perfectly clear. Current industry effort is tightly focused on building out all the standards-based Internet protocols that will be required for full voice over IP service and on creating both equipment and systems of "five 9's" (0.99999 availability and capability) robust-

---

[4]Armstrong, C. Michael, Chairman and CEO, AT&T Corporation, "Plain Talk about the Future," remarks delivered to the meeting "Internet World" in New York, October 8, 1998. Available online at <http://www.att.com/speeches/98/981008.maa.html>.

ness so that they can be brought into full service as soon as possible. The next-generation, voice over IP-based global telephone system is now well into its deployment phase all over the world.[5]

The committee submits, therefore, that the Navy will have little or no trouble acquiring Internet-based communications equipment that provides QOS guarantees sufficient for the tactical tasks at hand—namely, high availability, low loss, low delay, and prioritized traffic.

### C.2.3 Wireless Bandwidth

Contemporary wireless technology can provide orders-of-magnitude improvements in throughput over today's tactical radio systems. But this is only half the story. More important is that wireless data communication is an extremely "hot" area and that the technology is advancing by leaps and bounds, indeed, at Internet speeds. Just as is seen with fiber-optic transmission and switching technology, it is highly likely that data rates provided across wireless channels will grow geometrically over the near term in response to Internet demand. RF channels, of course, provide nothing like the potential bandwidth of fiber, and so wireless speeds will probably never come close to those available across fiber, but even the existing wireless technology can provide major advantages for the Navy.

The commercial wireless world is extremely fragmented, so it is impossible to provide a comprehensive overview of the field. Instead, three representative, wide-area systems are concentrated on here. Each occupies a very different point in the technology space and so the systems are quite different, with each being built by a major equipment vendor. The intent here is to show that the Navy already has a rather broad set of high-speed wireless technologies that it could choose from, if it so wished, and that each of these technologies is currently backed by a large and reputable manufacturer.[6]

- *Qualcomm high-data-rate technology.* This evolutionary advance in code division multiple access cellular technology provides air link speeds of up to 2.4 Mbps in a 1.25-MHz channel. It is an Internet-based technology that can be

[5]As one concrete example, Genuity reports that it was delivering over 100 million minutes of use per month in August 2000 on a QOS-enabled VOIP network that could at that time handle 80,000 concurrent phone calls.

[6]Manufacturer-supplied details for these systems may be found at the following Web sites: QUALCOMM Incorporated (San Diego, Calif.), High Data Rate System (HDR), <http://www.qualcomm.com/hdr/>; Cisco Systems, Inc. (San Jose, Calif.), WT-2700 Broadband System, <http://www.cisco.com/warp/public/cc/pd/witc/wt2700/>; and Terabeam (Seattle, Wash.), Fiberless Optical System, <http://www.terabeam.com>.

embedded in handsets, laptops, notebooks, and many other sorts of fixed or mobile devices.

- *Cisco WT2700 Suite.* This is a point-to-point, non-line-of-sight microwave radio system that provides speeds of up to 44 Mbps full duplex at ranges up to 30 miles within channel bandwidths of up to 12 MHz at about 2.5 GHz. It employs advanced modulation techniques such as vector orthogonal frequency division multiplexing and spatial and frequency diversity to take advantage of multipath signal reflections.
- *Terabeam free-space optical technology.* Terabeam is a Lucent-funded \$550 million venture that provides a high-speed (up to 1,000 Mbps) Internet service across 1,550-nanometer free-space optical links arranged into small hub-and-spoke cells. As is typical with optical solutions, the links can be very adversely affected by weather and indeed blocked altogether. However, field trials apparently indicate that reliable service may be possible at distances up to 1 km, even in cities such as Seattle.

Of course it could well be that none of the new technologies listed above turn out to be precisely suitable for the Navy's BMC3 wireless connectivity. However, they are all indicative of the technological revolution that is roiling the commercial wireless community. Going further, it seems highly likely to the committee that the Navy could benefit very significantly from applying some of this new technology to meet its wireless connectivity needs. In general, contemporary wireless technology provides very high bandwidth in an open, readily adaptable, standards-based package.

### C.2.4 Flexibility

With respect to flexibility, the commercial technology beats military systems hands down. Military radio systems, such as JTIDS (Link 16), are notorious for the extraordinarily detailed and voluminous planning that is required before they can be used. Entire staffs are devoted to planning tactical networks, and these plans often take months to prepare. This is a key weakness of such systems. It is so difficult to prepare radio plans that tactical operations may indeed suffer because the radio networks cannot be properly replanned fast enough to meet an evolving situation.

The situation is very different for commercial wireless technologies. Although certain types of wireless systems are indeed quite hard to plan—cellular base station planning comes to mind as an obvious example—most of the commercial technologies are designed so that they can be set up and brought into use almost immediately, by operators with relatively little specialized knowledge. Cellular phones are one case in point; when a subscriber acquires a cell phone, it is mandatory that this new, uninitialized phone be brought into the cellular provider's network as quickly and easily as possible. Point-to-point radio links are another case in point. Here the goal is to allow untrained purchasers to set up their own radio links within minutes after opening the packing cartons.

It is understood that the military operates under a number of restrictions on its use of RF spectrum and that these restrictions can complicate the planning and deployment of wireless networks. It is important to realize, though, that the commercial world operates under restrictions nearly as onerous. A great many of the Navy's planning and configuration problems are simply self-imposed (e.g., time-slot planning for JTIDS networks), and one can reasonably expect that commercial technology would be far simpler and more flexible than that of existing tactical radio systems.

### C.2.5 Military-specific Characteristics

Last but not least, a tactical communications system imposes certain requirements that are either unique to the military or far more stringent than their commercial analogs. Obvious examples include the ability to continue functioning in the presence of jamming (antijam) and low probabilities of interception or detection.

In general, commercial equipment is engineered without significant effort in these areas and hence cannot be directly employed in adverse tactical environments. On the other hand, some types of commercial wireless equipment inherently provide certain capabilities in this area, almost by accident as it were. For instance, point-to-point, free-air communications—and in particular optical links—are generally somewhat difficult to jam, unless by interposed obscurants, because they are highly directional. Similarly, commercial spread-spectrum systems offer a modest degree of protection against jamming and indeed somewhat lower the probabilities of detection or interception. It is conceivable that these levels of protection may prove adequate in some tactical scenarios. By and large, though, unmodified commercial technology is not suitable for tactical uses.

Perhaps surprisingly, commercial equipment performs particularly well in encryption and information assurance. Many vendors can supply wireless equipment that supports both link encryption and end-to-end data encryption. The commercially supplied encryption mechanisms are in general reasonably good and can often be readily replaced or augmented with military-grade encryption mechanisms as needed.

On the whole, then, the Navy should expect to devote resources to satisfying the purely military needs in wireless communications. However, existing commercial equipment often provides an excellent starting point for these modifications. In general, the Navy would be best served by adapting current state-of-the-art commercial wireless equipment to meet its tactical needs rather than engineering entirely new systems.

# D

# Abbreviated Description of U.S. Navy Short-Range Missile Defense Weapon Systems

## D.1 PHALANX CLOSE-IN WEAPON SYSTEM (CIWS)

- Low-cost, inner-layer self-defense system,
- Proposed in 1968, IOC in 1980; now on almost every combatant,
- Original system, completely autonomous; now integrated into SSDS and some Aegis weapon systems,
- Ku-band radar, closed-loop tracking of targets and bullets,
- 6-barrel gattling gun, 4500 rounds per minute, 500-round burst,
- 5-burst magazine, 1.1-km/s muzzle velocity,
- 20-mm depleted uranium shells, replaced with tungsten,
- Deliberate dispersion of shells to compensate for pointing errors,
- Range 5 km to 300 m, hit-to-kill,
- Effectiveness low outside a few kilometers and inside 0.5 km,
- Warhead detonation is the only effective short-range kill mechanism,
- Limited capability against maneuvering targets,
- Has engaged Exocets and Harpoons successfully in exercise demonstrations,
- Many versions in fleet with various upgrades,
- Block IB upgrade provided capability against surface targets, helicopters, and aircraft,
- Forward-looking infrared electro-optical imager/tracker, man-in-the-loop, and
- In the mid-1990s, was replaced with RAM in capital ships.

## D.2 SEA SPARROW MISSILE SYSTEM (RIM-7)

- Concept study in 1960s, using then-current ASCMs and aircraft as the threat,
- RIM-7H program begun about 1970 with multinational group (4 countries, grew to 13),
- Became the standard NATO ship self-defense missile,
- RIM-7M (Block II) introduced in 1978,
- RIM-7P programmable computer introduced in 1990,
- Weapon system concept consists of three elements: L-band radar + IFF + Mk-23 target acquisition system,
- Automatic fire control system (manned only to intervene in automatic process); Mk-57,
- Current sea sparrow missile (RIM-7P), based on Navy AIM-7F sparrow air-to-air missile,
- Same booster, guidance, and control with remote arming and "homing all the way" guidance,
- 20-cm (8-in.) diameter, 3.65 m long, 1-m wingspan,
- 350 m/s average speed, boost + coast,
- Motor-boost (3 seconds) + sustain (15 seconds),
- Optimum intercept range 1.5 to 6.5 km (6 to 25 s flight time),
- Weight 232 kg,
- Semiactive RF monopulse seeker,
- Target continuously illuminated by ship's radar,
- Warhead weighs 35 kg and is blast fragmenting,
- RF proximity fuse,
- Now on CVNs, LHDs, AOEs, AORs, DD963s,
- Not on older amphibious ships, and
- Has hit nonmaneuvering Styx and Exocet missiles in exercises.

## D.3 ROLLING AIRFRAME MISSILE (RAM) MK-31 GUIDED MISSILE SYSTEM

- Concept developed at Johns Hopkins University/Applied Physics Laboratory in the early 1970s to counter RF-emitting ASCMs,
- Missile dimensions: 12-cm (5-in.) diameter, 2.8 m long, 45-cm wingspan,
- 2 steerable canards and 4 tailfins for roll control,
- Missile weight: 74 kg,
- Employs Sidewinder solid rocket motor (Mk 36-8 or Mk 112),
- Boost (5 s) + glide,
- Optimum intercept range: 0.8 to 5 km (3 to 10 s),
- Launcher: 21-cell Mk-49,

- Launch rate: one every 3 s,
- Original seeker was "fire and forget," using passive, dual-mode (RF/IR) stinger missile seeker,
- Acquisition and initial track using RF from incoming missile,
- Transfer to IR for terminal-track, reticle scan in 4.1- to 4.5-mm band,
- Block I upgrade (RAM II) uses a linear-array IR detector in seeker,
- Dual-mode RF/IR or IR only,
- Completed operation tests, entered fleet in 1999,
- Maneuverable up to about 25 *g*,
- Engaged a maneuvering Vandal missile during an exercise,
- 9.5-kg warhead with 3-kg high explosive,
- New low-altitude fuze,
- Possible RAM upgrades,
- Dual-thrust motor,
- Larger diameter (14 cm to 15.5 cm), greater range,
- Maintains high velocity over entire trajectory,
- Increased maneuverability,
- Uplink to missile will allow target acquisition in bad weather,
- Sea RAM,
- Industry development in response to British navy request,
- Shorter inner range, faster response time,
- Uses phalanx 1B's high-resolution, target-search-and-track sensor, and
- CIWS 20-mm gun is replaced with RAM Block I, 11-tube launcher.

# E

# Biographies of Committee Members

**Alan Berman** is currently a part-time employee of the Center for Naval Analyses, where he assists with analyses of Navy R&D investment programs, space operation capabilities, and information operations. He also consults for the Applied Research Laboratory of Pennsylvania State University (ARL/PSU), where he provides general management support and program appraisal. Dr. Berman's background is in defense science and technology, particularly in regard to advanced weapon and combat systems. He is regarded as a leading expert on combat systems. Positions he has held include, among others, dean of the Rosenstiel School of Marine and Atmospheric Sciences at the University of Miami and director of research at the Naval Research Laboratory. Dr. Berman has served on numerous government advisory and scientific boards. He is currently a member of the Naval Studies Board (NSB). He is also a member of the Free Electron Laser (FEL) oversight board, which advises Jefferson National Laboratory of the Department of Energy on its FEL program.

**Roy R. Buehler** is retired from Lockheed Martin Aeronautical Systems, where he managed Lockheed Martin's U.S. Customs Service line of business. Mr. Buehler's background is in antiair and antisurface warfare and airborne early warning systems. He has more than 30 years of experience in industry and government as an experimental test pilot, business planner, and program manager in the start-up of new aircraft programs such as the F-111, F-14, F-18, A-6, and F-22/Naval Advanced Tactical Fighter. He served in the Navy both as a carrier fighter pilot and as an experimental test pilot and major program manager. Mr. Buehler is a member of the Society of Experimental Test Pilots.

**William A. Davis, Jr.**, an independent consultant, retired as deputy program manager for ballistic missile defense (BMD) in 1982 following 33 years of government service. Today Mr. Davis serves as an independent consultant on technology matters relating to national missile defense (NMD) and BMD. His recent clients include the Joint Ballistic Missile Defense Office and the Army NMD Program Office. Mr. Davis's background is in tactical and strategic missile defense, as well as missile research and development. Upon retirement from the government, Mr. Davis served as vice president for space defense at Teledyne Brown Engineering, where he directed simulation and analysis of tactical warning/attack assessment network and space-based elements of the Strategic Defense Initiative. Mr. Davis has served on numerous government advisory and scientific boards, including Department of Defense task forces on tactical missile research and development.

**Larry E. Druffel** is president of South Carolina Research Authority, a public, nonprofit organization engaged in applying advanced technology to increase industrial competitiveness. Previously, he was director of the Software Engineering Institute and served as the vice president for business development at Rational Software. Earlier in his career, Dr. Druffel was on the faculty at the U.S. Air Force Academy. He later managed research programs in advanced software technology at DARPA, was founding director of the ADA Joint Program Office, and then served as director of computer systems and software (Research and Advanced Technology) in the Office of the Secretary of Defense. He is a fellow of the Institute of Electrical and Electronics Engineers (IEEE) and the Association for Computing Machinery (ACM).

**Brig "Chip" Elliott** is principal scientist at BBN Technologies. Mr. Elliott's background is in Internet and wireless network technologies, tactical communications systems, and space-based surveillance and communications. As the technical lead scientist at BBN, he uses Internet technology to build networks for international corporations and U.S. government agencies. He was the chief architect for the networking component of the Army's Near-Term Digital Radio Program, which forms the backbone of the Army's Tactical Internet; for the British Army's High Capacity Data Radio network; and for the Canadian Army's IRIS network. He has also acted as lead for a number of LEO satellite systems (Discoverer II, SBIRS-low, Celestri) as well as a proposed undersea network.

**Frank A. Horrigan** retired from the Technical Development Staff for Sensors and Electronic Systems at Raytheon Systems Company. Dr. Horrigan has a broad general knowledge of all technologies relevant to military systems. A theoretical physicist by training, he has more than 35 years of experience in advanced electronics, electro-optics, radar and sensor technologies, and advanced information systems. In addition, he has extensive experience in planning and managing IR&D investments and in projecting future technology growth directions. Dr. Horrigan once served as a NATO fellow at the Saclay Nuclear Research Center in France. Today he serves on numerous scientific boards and advisory committees, including the NRC's Army Research Laboratory Technical Assessment Board and Naval Studies Board, and he recently served on the Committee for the Review of ONR's Uninhabited Combat Air Vehicles (as its chair).

**Richard J. Ivanetich** is director of the Computer and Software Engineering Division at the Institute for Defense Analyses (IDA). Dr. Ivanetich has extensive experience in missile and defense systems technology, particularly in regard to the system architecture design of information systems relating to strategic and theater nuclear forces. Before becoming director, Dr. Ivanetich was assistant director of the Systems Evaluation Division at IDA. Before joining IDA, he was an assistant professor of physics at Harvard University. Dr. Ivanetich is a member of the Naval Studies Board.

**Harry W. Jenkins**, a retired major general in the U.S. Marine Corps, is director of business development and congressional liaison at ITT Industries-Defense, where he is responsible for activities in support of tactical communications systems and airborne electronic warfare between the Navy, Marine Corps, National Guard, and appropriate committees in Congress. General Jenkins' operational background is in expeditionary warfare, particularly in its mission use of C4I systems. During Desert Storm, General Jenkins served as the commanding general of the Fourth Marine Expeditionary Brigade, directing operational planning, training, and employment of the ground units, aviation assets, and command and control systems in the 17,000-man amphibious force. General Jenkins's last position before retirement from the Marine Corps was director of Expeditionary Warfare for the Chief of Naval Operations; as director, he initiated a detailed program for C4I systems improvements for large-deck amphibious ships and reorganized the Navy's UAV efforts for operations from aircraft carriers and amphibious ships. He is a member of numerous professional societies, including the Navy League and the Aerospace Industries Association.

**David V. Kalbaugh** is head of the Power Projection Systems Department at the Johns Hopkins University/Applied Physics Laboratory (JHU/APL), where he is responsible for the enhancement of U.S. military forces through the development and test of power projection (strike warfare) systems and technologies. Dr. Kalbaugh's background is in missile, precision strike, and tactical and strategic communication systems. He joined JHU/APL in 1969 and was involved in the development of the Tomahawk cruise missile system at its inception in the early 1970s. In addition to his supervisory and management duties, Dr. Kalbaugh taught for more than a decade in JHU's Whiting School of Engineering. He has served on numerous government and scientific advisory boards, including participation in tasks for the Undersecretary of Defense for Acquisition and the Program Executive Officer for Theater Air Defense.

**Frank Kendall** is an independent consultant with more than 25 years of engineering, management, and national security experience in both the private and government sectors. In government, he had assignments in the Department of Defense in systems engineering and acquisition management, including positions in the Office of the Secretary of Defense as the director of Tactical Warfare Programs and as Assistant Deputy Undersecretary of Defense for Strategic Defensive Systems. Mr. Kendall was formerly the vice president of engineering at the Raytheon Company and also spent 10 years on active duty with the Army, serving as an air defense missile unit commander and as assistant professor of engineering at West Point. He has served on numerous government and scientific advisory boards. He is a member of the Army Science Board and vice chairman of the Defense Intelligence Agency S&T Board, and he has served as a consultant on several Defense Science Board studies.

**L. David Montague,** an independent consultant, is retired president of the Missile Systems Division at Lockheed Martin Missiles and Space. Mr. Montague, a member of the National Academy of Engineering (NAE), has a strong background in military weapon systems, particularly in regard to guidance and control of submarine-launched weaponry. His experience has focused on both tactical and strategic strike systems as well as on the requirements for, development of, and policy issues related to defense systems to protect against weapons of mass destruction. His recent research interests include electric vehicles powered by battery or fuel cells integrated with induction-drive, high-speed highway automatic headway and vehicle control. Mr. Montague is a fellow of the American Institute of Aeronautics and Astronautics and has served on numerous government and scientific advisory boards, including task forces for both the Army and the Defense Science Board.

**F. Robert Naka** is president and CEO of CERA, Inc. Dr. Naka, a member of the NAE, has a strong background in reconnaissance, surveillance, and communication and command systems, components, and technologies. He is widely known for his contributions to the development of National security systems and his contributions in materials and sensor technologies for advanced military systems. Dr. Naka's career has spanned a broad range of assignments, including research and teaching duties at the university level and serving as the Air Force's chief scientist. Dr. Naka has served on numerous government and scientific advisory boards, including the NRC's Air Force Studies Board, in 1972-1975 and 1978-1982, and, recently, the Committee on Counterforce Options Against Tactical Missile Systems.

**J. Theodore Parker,** retired vice admiral, U.S. Navy, is an independent consultant. Admiral Parker's background is in research and development and in military operations relating to ship air defense systems. In his operational tours as flag officer, Admiral Parker commanded Service Group Two, the Mobile Logistics Support Force for the Atlantic Fleet. Additionally, he commanded the Operational and Test and Evaluation Force that tested new weapon system designs for the Navy. Before retiring in 1989, Admiral Parker served as director of the Defense Nuclear Agency.

**C. Kumar N. Patel** is the former vice chancellor of research at the University of California at Los Angeles, where he is now serving as a professor. Until 1993, Dr. Patel served as executive director of the Research, Materials Science, Engineering, and Academic Affairs Division at AT&T Bell Laboratories. Dr. Patel, a member of the NAE and NAS, has an extensive background in several fields, including materials, lasers, and electro-optical devices. During his career at AT&T, which began in 1961, he made numerous seminal contributions in several fields, including gas lasers, nonlinear optics, molecular spectroscopy, pollution detection, and laser surgery. Dr. Patel has served on numerous government and scientific advisory boards. He is past president of Sigma Xi and of the American Physical Society. In addition, Dr. Patel has received numerous honors, including the National Medal of Science, for his invention of the carbon dioxide laser.

**Nils R. Sandell, Jr.,** is president and CEO of ALPHATECH, Incorporated. Dr. Sandell's background includes automatic target recognition and sensor management technologies, as well as algorithms for airborne reconnaissance and tracking and sensor resource management for ground moving target indicator and synthetic aperture radar. At ALPHATECH, he is currently responsible for projects that are developing, planning, and scheduling algorithms for airborne reconnaissance platforms. He is a former associate professor at the Massachusetts Institute of Technology, where he lectured in the areas of estimation and control theory, stochastic processes, and computer systems.

**Howard E. Shrobe** is associate director and principal research scientist of the Massachusetts Institute of Technology's Artificial Intelligence Laboratory (MIT AII). Dr. Shrobe's research is in intelligent systems, particularly in regard to knowledge-based software development. From 1994 to 1997, Dr. Shrobe served as assistant director and chief scientist of the DARPA Information Technology Office, where he was responsible for the Evolutionary Design of Complex Software and Information Survivability programs. At MIT AII, Dr. Shrobe's research efforts include knowledge-based collaboration webs, dynamic domain architecture, and intelligent information infrastructure projects.

**John P. Stenbit** is executive vice president of TRW, Inc. Mr. Stenbit, a member of the NAE, has an extensive background in missile and space systems, communication systems and networks for military systems, and computer networking and communications. His interests have focused on system engineering problems in which boundary conditions are variable and have nonlinear distortions caused by regulation, treaty, or perhaps technological change, such as those associated with strategic offensive and defensive missiles in the face of arms control treaties. Mr. Stenbit has served on numerous government and scientific advisory boards.

**Robert F. Stengel** is professor of mechanical and aerospace engineering and director of the Laboratory for Control and Automation at Princeton University. His current research focuses on failure-tolerant and robust control, intelligent systems, and the coordinated flight of uninhabited air vehicles. At Princeton, Dr. Stengel was director of the Flight Research Laboratory, where he conducted pioneering experimental research on digital flight control systems, flight computer networking via fiber optics, aircraft flying qualities, and aerodynamic system identification. Before coming to Princeton, Dr. Stengel held positions with the Analytic Sciences Corporation, the Charles Stark Draper Laboratory, the U.S. Air Force, and NASA. He is a fellow of the American Institute of Aeronautics and Astronautics and of the Institute of Electrical and Electronics Engineers. He received the AIAA Mechanics and Control of Flight Award in 2000.

**Edward J. Wegman** is professor and director of the Center for Computational Statistics at George Mason University (GMU). Dr. Wegman came to GMU with an extensive background in both theoretical statistics and computing technology. His early career was spent as an assistant professor at the University of North Carolina's Department of Statistics and as head of the Mathematical Sciences Division at the Office of Naval Research. Additionally, Dr. Wegman was the original program director of the basic research program in ultrahigh-speed computing at the Strategic Defense Initiative Innovative Science and Technology Program Office (Star Wars Program). Dr. Wegman is a fellow of the American Statistical Association, the American Association for the Advancement of Science, and the Institute of Mathematical Statistics. He has served on numerous government and scientific advisory boards and is currently a member of the NRC's Panel on Survivability and Lethality Analysis.

**Stephen D. Weiner** is a senior staff member in the Systems and Analysis Group at the Lincoln Laboratory of the Massachusetts Institute of Technology (MIT). Dr. Weiner's background in ballistic missile defense includes system and radar design, sensor tracking and discrimination measurements, and interceptor guidance. His research interests also include defense against both theater and strategic cruise missiles. Dr. Weiner has served on a number of government and scientific advisory panels, including a 1991 naval research advisory committee on naval theater ballistic missile defense.

# F

# Agendas for Meetings of the Committee for Naval Forces' Capability for Theater Missile Defense

**APRIL 25-26. 2000**
**NATIONAL RESEARCH COUNCIL, WASHINGTON, D.C.**

**Tuesday, April 25, 2000**

**Closed Session: Committee Members and NRC Staff Only**

0830 CONVENE—Welcome, Composition and Balance Discussion, Report Preparation
Dr. Alan Berman, Committee Chair
Dr. Myron F. Uman, CPSMA Director
Dr. Ronald Taylor, NSB Director

1300 NAVY THEATER MISSILE DEFENSE—Surface Warfare Requirements and Programs
CAPT Michael Moe, USN, Deputy Director, Theater Air Warfare (N865B)

1500 MARINE CORPS THEATER MISSILE DEFENSE—Operational Requirements
Col William L. Groves, USMC HQMC Aviation Command and Control Programs

1900 END OF SESSION

**Wednesday, April 26, 2000**

**Data-Gathering Meeting Not Open to the Public: Classified Discussion (Secret)**

0815 CONVENE—Welcome, Discussion
Dr. Alan Berman, Committee Chair
Dr. Charles F. Draper, Senior Program Officer

0830 DEPARTMENT OF THE NAVY S&T—Naval Expeditionary Warfare S&T
Dr. Eli Zimet, Head, Naval Expeditionary Warfare S&T and Special Programs Department, Office of Naval Research (Code 35)

1030 DEPARTMENT OF THE NAVY S&T—Theater Missile Defense S&T
CAPT A.J. Cetel, USN, Program Manager, Theater Missile Defense S&T Programs, Office of Naval Research (Code 35)/ Office of the Chief of Naval Operations (N091)

1300 FUTURE NAVAL CAPABILITIES RELATED TO THEATER MISSILE DEFENSE—Missile Defense, Platform Protection
Dr. Joseph P. Lawrence III, Naval Research Laboratory

**Closed Session: Committee Members and NRC Staff Only**

1430 COMMITTEE DISCUSSION—Report Preparation

1700 ADJOURN

## MAY 23-24. 2000
## NATIONAL RESEARCH COUNCIL, WASHINGTON, D.C.

### Tuesday, May 23, 2000

**Closed Session: Committee Members and NRC Staff Only**

0830 CONVENE—Welcome, Report Discussion
Dr. Alan Berman, Committee Chair
Dr. Charles F. Draper, Senior Program Officer

**Data-Gathering Meeting Not Open to the Public: Classified Discussion (Secret)**

0900 BALLISTIC MISSILE DEFENSE ORGANIZATION (BMDO)—Programs, Technologies, Operations
COL Robert Barnes, USA, Team Leader, Upper Tier Program Support Team, BMDO
CDR Swicker, USN, Upper Tier Program Support Team, BMDO

1045 CENTRAL INTELLIGENCE AGENCY (CIA)—Theater Missile Threats, Land Attack Cruise Missiles (LACMs), Antiship Cruise Missiles (ASCMs)
Mr. Scott E. Hopkins, Mr. Joseph M. Irek, Mr. Robert C. Merkel, CIA

1300 DEFENSE INTELLIGENCE AGENCY (DIA)—Theater Missile Threats, Theater Ballistic Missiles (TBMs)
Maj Stephen A. Williams, USAF, DIA

1400 Joint Theater Air Missile Defense Organization (JTAMDO)—Programs, Technologies, and Operational Requirements
CAPT John McMurtrie, USN, JTAMDO

1500 Air Force Airborne Laser (ABL)
Maj Gary Henry, USAF, Air Force Global Power Program
Dr. William E. Thompson, Directed Energy Directorate, Air Force Research Laboratory (AFRL)

1600 Theater Missile Threats
Dr. Martin H. Lindsey, Intelligence Officer, Missile Intelligence and Space Center, DIA

1700 End of Session

**Wednesday, May 24, 2000**

**Closed Session: Committee Members and NRC Staff Only**

0815 Convene—Welcome, Introductory Remarks, Report Discussion
Dr. Alan Berman, Committee Chair
Dr. Charles F. Draper, Senior Program Officer

**Data-Gathering Meeting Not Open to the Public: Classified Discussion (Secret)**

0830 Navy Theater Missile Defense
RADM Rodney P. Rempt, USN, Deputy Assistant Secretary of the Navy for Theater Combat Systems

1000 ABM Treaty Limitations
CAPT Mark Rosen, USN (retired), Center for Naval Analyses

1115 Theater Air and Missile Defense—Aegis Weapon System Technology Upgrades
CAPT O.H. Perry, USN, Program Manager, Future Theater Air and Missile Defense, Naval Sea Systems Command, PMS 456

1315 Cooperative Engagement Capability—Remote Cueing, Network Architecture
Mr. Richard Johnson, Technical Director, Cooperative Engagement Capability, Program Executive Office for Theater Surface Combatants (PEO TSC)

1415 Army Air and Missile Defense—Programs, Technologies, Operations
Dr. Shelba Proffitt, Deputy Program Executive Office for Air and Missile Defense (PEO AMD)
Mr. Tony Cosby, Chief Engineer, PEO AMD

**Closed Session: Committee Members and NRC Staff Only**

1530 Committee Discussion—Report Discussion
1700 Adjourn

## JUNE 27-28, 2000
## NATIONAL RESEARCH COUNCIL, WASHINGTON, D.C.

### Tuesday, June 27, 2000

**Closed Session: Committee Members and NRC Staff Only**

0815 Convene—Welcome, Report Deliberation
Dr. Alan Berman, Committee Chair
Dr. Charles F. Draper, Senior Program Officer

**Data-Gathering Meeting Not Open to the Public: Classified Discussion (Secret)**

0915 Program Executive Office Theater for Surface Combatants (PEO TSC)—Overview
RADM William W. Cobb, Jr., USN, Program Executive Officer for Theater Surface Combatants
0930 Navy Area TBMD
Mr. Jerry Lacamera, Program Manager, Navy Area TBMD (PMS 451)
CDR Alan E. Haggerty, USN, PEO TSC
Mr. William Ainsley III, Technology Service Corporation
Mr. Harry D. Farley, PEO, Theater Air Defense (PMS 422B)
Mr. Joel D. Miller, APL, Johns Hopkins University
1315 Ship/Area Antiair Warfare Defense
Mr. William S. Smothers, Deputy Program Manager, CSE&I, PEO for Expeditionary Warfare
Mr. Joseph F. Williams, SPY-1 Radar System Engineer, PEO TSC
1530 MFR/DD21
Mr. Am P. Supsiri (PMS 500)
Mr. Gabriel Moskovitz, Raytheon Electronic Systems
1600 Discussion Moderator: Dr. Alan Berman, Committee Chair
1700 End of Session

### Wednesday, June 28, 2000

**Closed Session: Committee Members and NRC Staff Only**

0815 CONVENE—Welcome, Introductory Remarks, Report Discussion
Dr. Alan Berman, Committee Chair
Dr. Charles F. Draper, Senior Program Officer

**Data-Gathering Meeting Not Open to the Public: Classified Discussion (Secret)**

0830 NAVY THEATER WIDE TBMD
CDR Sheila Patterson, USN, PEO TSC

1100 BALLISTIC MISSILE DEFENSE SYSTEM ENGINEERING
Mr. Richard Ritter, Deputy, System Engineering, Ballistic Missile Defense Organization

1330 BATTLE MANAGEMENT C4I
Mr. Eugene Warner, PEO TSC

1500 OVERLAND CRUISE MISSILE DEFENSE
Mr. Larry Lefbom, PEO TSC
Mr. David A. Bement, APL, Johns Hopkins University
CDR Sheila Patterson, USN, PEO TSC

**Closed Session: Committee Members and NRC Staff Only**

1530 COMMITTEE DISCUSSION—Report Deliberation

1700 ADJOURN

## JULY 25-26, 2000
## NATIONAL RESEARCH COUNCIL, WASHINGTON, D.C.

### Tuesday, July 25, 2000

**Closed Session: Committee Members and NRC Staff Only**

0830 CONVENE—Welcome, Report Discussion
Dr. Alan Berman, Committee Chair
Dr. Charles F. Draper, Senior Program Officer

**Data-Gathering Meeting Not Open to the Public: Classified Discussion (Secret)**

0900 THEATER AIR AND MISSILE DEFENSE—Engineering the System
RADM David M. Altwegg, USN (retired), Deputy Director,

Theater Air and Missile Defense and Systems Engineering, PEO TSC

1045 JOINT MISSILE DEFENSE TECHNOLOGY PROGRAM—The Prometheus Study; Missile Convolution Engine and Mass Moment Missile Technology
Mr. Mick L. Blackledge, Senior Engineer for Missile System Testing, Joint Technology Program

1245 AREA DEFENSE—Concepts of Operations
CDR Michael Delaney, USN, Naval Sea Systems Command (PMS 467)

1430 TACTICAL AIRCRAFT IN THEATER MISSILE DEFENSE—E-2C Role and Improvements
LCDR Ramon A. Collazo, Jr., USN, PEO Tactical Aircraft Programs

1600 ROLE OF ELECTRONIC WARFARE IN AREA/SELF-DEFENSE
CAPT Deborah R. Stiltner, USN, Electronic Warfare System Program Manager, PEO TSC
Dr. Charles Heider, Head, Electronic Warfare Special Program Office, Naval Research Laboratory, Code 5700

1730 END OF SESSION

**Wednesday, July 26, 2000**

**Closed Session: Committee Members and NRC Staff Only**

0830 CONVENE—Welcome, Introductory Remarks, Report Discussion
Dr. Alan Berman, Committee Chair
Dr. Charles F. Draper, Senior Program Officer

**Data-Gathering Meeting Not Open to the Public: Classified Discussion (Secret)**

0845 AIM 120 ADVANCED MEDIUM-RANGE AIR-TO-AIR MISSILE (AMRAAM)
LCDR Robert Vance, USN, Air Warfare Division, Office of the Chief of Naval Operations (N880C7)

**Closed Session: Committee Members and NRC Staff Only**

1015 COMMITTEE DISCUSSION—Report Discussion
Moderator: Dr. Alan Berman, Committee Chair

**Data-Gathering Meeting Not Open to the Public: Classified Discussion (Secret)**

1300 LINK 16
Dr. Kenneth L. McCloud, Advanced Tactical Data Links Systems Program Office, Space and Naval Warfare Systems Command (PMW 159)

1530 TACTICAL PLATFORMS IN THEATER MISSILE DEFENSE—Joint Land Attack Cruise Missile Defense Elevated Netted Sensors (JLENS) Role
COL Mary Fuller, USA, Program Manager, JLENS Program Office

1700 ADJOURN

## AUGUST 29-30, 2000
## NATIONAL RESEARCH COUNCIL, WASHINGTON, D.C.

### Tuesday, August 29, 2000

**Closed Session: Committee Members and NRC Staff Only**

0830 CONVENE—Welcome, Opening Remarks, Report Deliberation
Dr. Alan Berman, Committee Chair
Dr. Charles F. Draper, Senior Program Officer

**Data-Gathering Meeting Not Open to the Public: Classified Discussion (Secret)**

0900 Marine Corps Theater Missile Defense
LtGen Fred McCorkle, USMC, Deputy Commandant for Aviation, Headquarters, Marine Corps
Col William L. Groves, USMC, Aviation Command and Control Programs, Headquarters, Marine Corps

1300 NAVY THEATER MISSILE DEFENSE—Requirements, Programs, Concepts of Operation, Deconfliction Issues
RADM John M. Kelly, USN, Director, Theater Air Warfare, Office of the Chief of Naval Operations (N865)

1430 DEPARTMENT OF DEFENSE AND DEPARTMENT OF THE NAVY MISSILE SYSTEMS DEVELOPMENT—Overview, Future Capabilities, and Other Issues
RADM Wayne E. Meyer, USN (retired), W.E. Meyer Corporation

1545 THEATER MISSILE DEFENSE—Capstone Requirements Document, Other Discussion
CAPT Ward Clark, USN (retired), Theater Air and Missile Defense, United States Joint Forces Command, J85 Directorate
1700 END OF SESSION

**Wednesday, August 30, 2000**

**Closed Session: Committee Members and NRC Staff Only**

0830 CONVENE—Welcome, Introductory Remarks, Report Deliberation
Dr. Alan Berman, Committee Chair
Dr. Charles F. Draper, Senior Program Officer

**Data-Gathering Meeting Not Open to the Public: Classified Discussion (Secret)**

0900 ELECTRO-OPTICAL TECHNOLOGY—Focal Plane Arrays, Lasers, EO Systems
Dr. Karl A. Harris, Director, Electro-Optics Center, Applied Research Laboratory, Pennsylvania State University
1045 TACTICAL HIGH-ENERGY LASER SYSTEMS—Israeli System, Army Programs
Mr. Richard J. Bradshaw, Program Manager, Directed Energy Technology Programs, Army Space and Missile Defense Command
1215 AIRBORNE TACTICAL LASER
Mr. Kevin Stull, Joint Non-Lethal Weapons Directorate
Mr. Donald C. Slater, The Boeing Company
1300 THE VALUE OF THE SINGLE INTEGRATED AIR PICTURE
Dr. Joan F. Cartier, Science and Technology Division, Institute for Defense Analyses
Dr. Jeffrey F. Nicoll, Science and Technology Division, Institute for Defense Analyses
1430 SINGLE INTEGRATED AIR PICTURE—System Engineering Efforts
CAPT Jeffrey W. Wilson, USN, SIAP System Engineer, Naval Sea Systems Command

**Closed Session: Committee Members and NRC Staff Only**

1600 COMMITTEE DISCUSSION—Report Deliberation
Moderator: Dr. Alan Berman, Committee Chair
1700 ADJOURN

## SEPTEMBER 11-15, 2000
## J. ERIK JONSSON WOODS HOLE CENTER, WOODS HOLE, MASSACHUSETTS

**Closed Session: Committee Members and NRC Staff Only**

0830 CONVENE—Welcome, Administrative Issues, Meeting Schedule
Dr. Alan Berman, Committee Chair
Dr. Charles F. Draper, Senior Program Officer

MONDAY, SEPTEMBER 11
TUESDAY, SEPTEMBER 12
WEDNESDAY, SEPTEMBER 13
THURSDAY, SEPTEMBER 14
FRIDAY, SEPTEMBER 15

0845 WORKING SESSION—Committee Deliberations, Report Writing
1300 WORKING SESSION—Committee Deliberations, Report Writing
1700 END OF SESSION

## NOVEMBER 28-29, 2000
## NATIONAL RESEARCH COUNCIL, WASHINGTON, D.C.

**Closed Session: Committee Members and NRC Staff Only**

0800 CONVENE—Welcome, Administrative Issues, Meeting Schedule
Dr. Alan Berman, Committee Chair
Dr. Charles F. Draper, Senior Program Officer

TUESDAY, NOVEMBER 28
WEDNESDAY, NOVEMBER 29

0815 WORKING SESSION—Committee Deliberations, Report Writing
1300 WORKING SESSION—Committee Deliberations, Report Writing
1700 END OF SESSION